Aliens Among Us

Jose Vasquez

Introduction

My book contains information I gathered from my dreams and based on my theories of alien life and how they had a huge impact on our lives by helping us evolve at an alarming rate. By completing this task, they allowed us to be their slaves and work for the rest of our lives for their cause.

As you read further into my book, you will realize that my characters in my story have traveled a path that lead all of them to meet each other. Along their journey on completing their missions, they discover all the secrets that were kept by the government.

This will eventually haunt the human race, for as long as we live, for the first time, we will find the truth about Atlantis, aliens, cloning, Greek mythology, and technology that has not yet been created or thought of but has been mentioned in this book.

The biggest secrets

that have been kept from us are that the aliens have time-traveled to our past and evolved us at a rapid rate. They also made an agreement with our government to keep everything silent to keep us under their control.

The mighty Greek

Olympians were actually ancient astronauts, also known as aliens, from a different galaxy and planet. So brace yourself for a story for the ages. We will also discover planet Nubu, how their culture reacts to the human race, and aliens coexisting in the same planet as equals.

These aliens can't stand next to the site of a human being, and they're trying to keep their body temperature at 128 degrees Fahrenheit to maintain their existence. If they can't manage to keep their body temperature, drastic changes may occur as in death. They were also trying to raise earth's temperature for their benefit and total disregard for human life.

Characters

Members of the Alien Resistance Team

Robert Young

Steven Williams

Rugeey Watson

Amber Rivers

Michael Ross

Joseph Parker

Jayden Walker

Louis Ruth

Eric Stevens

Joey Montero

Pete Davis

David Jackson

Alex Sr

Luke Washington

USA Officials

George Washington

(First President of US)

Thomas Jefferson (Vice president of US)

James R. Hancock (CIA Agent)

Aliens

Alckamus (Head of the aliens)

Venum (Second in charge of the aliens)

Naboko (Head of alien security)

Dream Master (Controls the dream and underworld)

Characters on Planet Nubu

General Dex (Head of

the alien army)

Private Grunt (Regular alien soldier)

Helios (The aliens worship him)

Hanos (The aliens known as evil ones)

Theos (Well-known science and technology major)

On October 10, 1991, two young college students, physics and science and chemistry majors, by the names of Robert Young and Steven Williams created a time machine so they could go and travel back through time with the intent of changing our past, possibly altering our future, at about three million years ago but ended up on July 4, 1776, and came to find out the secrets about the United States government.

The two physics

majors found upon their discovery that the United States government had been controlled by an alien race (hybrid aliens) who were created by the aliens using their DNA and that of Neanderthals.

While combining the two DNA structures,

the vision of Steven and Roberts began to change dramatically. They now were able to see certain humans they once knew; they now looked like reptilian creatures.

Upon all their discoveries made, Steven started to tell people in the current time period that they were not actually humans. This stirred quite a huge reaction among the neighborhood, and at the same time, Robert and Steven sent a strong message to the government.

The president of the United States of

America heeded the message and gave a direct order to have Steven murdered. Robert started to feel a light bulb flash above his head that was to send Steven back in time to disrupt the control the alien race had over the human race.

Steven arrived back in the past on July 4, 1776. As Steven arrived, he noticed that he had been followed by what appeared to be a tall man, who was roughly thirteen feet tall. The man asked him, “What are you doing here? You don’t belong here. You will be destroyed if you continue to travel through time."

As Steven arrived back into the present time, he noticed the same man standing in front of him. He asked the man, "Sir, what is your name?"

"I'm Alckamus."

"I think you're making a huge mistake, trying

to stop us. We can't be stopped. We're an advanced race."

"My kind have been around for ten thousand years. We are also ahead of humans when it comes to technology and other aspects of life. We also created the human race. We're not made like you guys. Our insides are made of gold. Humans were designed as our slaves. How can you kill us? Tell me!”

In response, Steven answered Alckamus, “I’m going to try down to my very last breath.”

Alckamus vanished.

Steven contacted Robert through a special time-travel cellular phone that they both created and tested.

“Robert, they’re on to us and are trying to kill us.”

“Steven, we anticipated this would happen.”

"What should I do?"

"Go hide now somewhere where they might not look so this could allow us to buy us some time so we can think straight, man. Go get some rest. Okay?"

"Oh, I'm in a motel. I'm checkin' in."

Steven entered a

motel. “Hello, sir, do you have any rooms available?”

“Sorry the only rooms I have available are on the fourth floor. Room 420 is in that floor, and it’s good enough for you. It’s the only one that has one bed.”

“Okay, I’ll take it.”

As Steven sat on his

bed, he decided to call Robert.

"I got it. All you have to do is create an undetectable suit that allows us to go undetectable while we're in the presence of these aliens. That shape-shifts into any physical form along with making someone invisible to the aliens."

"Hey, Robert, you know, you're right. I should've thought about that, and while I'm at it, I could use the suit to go in our favor by learning all their weaknesses and strengths."

"Steven, remember, keep me posted on

how effective the suit is while testing. Two heads are better than one. Oh, by the way, how did the alien look? You know the one that said he was gonna kill you."

"Well, if you must know, he looked human with a long stretched head and was about thirteen feet long. He was rather strong looking. I'm afraid I have to let you go now. I will keep in touch."

Meanwhile, Robert was back in his apartment at Glendale Road in Houston Texas, trying to figure out a solution to overcome and overthrow these aliens as he wondered how it could affect or have a negative impact on every single form of life existed.

Robert thought, but something had to be done in order for the human race to be freed from slavery. That's it.

I'm going back in time to meet up with Steven so I could kick some alien tail. I'm going into my time machine and setting the coordinates, and I'm ready to rock and roll.

"Wow, that was fast."

"Hey, Steven, are you ready to get some action?"

"Hell yeah! Let's go to my room 420 and discuss what our next moves are going to be."

A loud noise came from the window. "Let's check it out. It's Alckamus, and he is waving at us."

"Hey, let's create a self-containment field

to draw and hold Alckamus."

"Great idea, Rob."

"It's all done, Steven. Check this out. How perfection is not even close to what I've created. It's a masterpiece. All that should be said is that we must learn this alien race and study them like they studied us for years and years. It's been two days, and we're almost done, Steven. The containment center looks great. Let's turn it on.

A switch turned on ten, seconds later. "Look, Rob, it's Alckamus. We caught him. It works, really works right on."

"Why are you trying to kill us, Alckamus?"

"'Cause you're a threat to our very existence. That's why I'm trying to kill you both. Our goal when you and your partner were created was you would be intelligent enough to serve our purpose and would be our slaves. What I did not intend was, you and your blood brother Steven would try and stop us."

"How did you create us, Alckamus. We must know."

"Your kind was created in a test tube.

Ha-ha, from my planet Nubu. We purposely locked 95 percent of your brains for a reason of keeping your brain waves stabilized at certain level. We need gold to survive, and your planet has a large supply of it, and we need it. As slaves, your race will bring us the gold. We control the economy and your precious government. We own basically the whole planet earth.

“If it wasn’t for us altering your DNA, your race would be as smart as apes as you know we’re close to being immortal, but we can die. Our life span is ten thousand years. I have spoken too much. That's it!

"I forgot to mention the aliens from my planet are coming to earth as we speak. Ha-ha!"

"No no! Robert, what do we do?"

"Steve, I'll make the weapons, and you are in charge of getting an army together. Hey, Steve, I bumped into Rugeey. She said, 'How's Steve? I miss him. I remember when we dated in high school, I used to mess with him about being a nerd, but the truth is I really liked nerds. I just didn't know it at the time. Tell him to call me. I need to talk to him. I just moved back into town. I decided not to reenlist back in the army. My number is still the same. Thanks, Rob.' Dude, you need to call her. I'm gonna start making the gold disintegrator grenade."

"I will call her after I'm done building all

my weapons, the relocator, containment gun, and hologram boomerang that makes the sharpest blade look dull. Each weapon serves a purpose during a war of this magnitude. The relocator gun not only sends someone somewhere else, it also freezes you at the same time and keeps you in that state for eternity."

"I'm also creating a hover sneaker, which allows you to fly if necessary. What do think, Steven?"

"Very good, Rob. I'm gonna call Rugeey. I'll be back tomorrow." Steven left the door closed. He arrived at his secret hideout. His parents pool house located at 10 Beaver Street.

"Hey, Rugeey, how's it going? It's Steven." "Hey, Steven, how have you been? It's been a long time. I know what brings you back. Well, I guess Robert told you what I told him."

"Yeah, he did."

"I just wanted to get to know you again. I know I wasn't a great girlfriend in high school. I want to make up for that."

"Sure, you can. Starting now, you're going to assemble the best fighters and you're gonna lead them into a battle against aliens which created the human race. Got a problem with that?"

"We're gonna need weapons and protective gear."

"Robert and I have you covered. Meet me at 10 Beaver Street at

9:00 a.m. with your guys. I'm gonna send you a portal that will allow you to meet me here in the past."

"Wow! Steven, you and Robert came along way. You guys are very smart."

"I know. Thanks. Well, I will see you guys tomorrow. It's 9:00

a.m., and Rugeey was passing through the portal with her ten men.

“Hey, Steven, I’m here, and these are the guys I have. Alex the Great, Amber, Michael, Joseph, Jayden Louis, Eric, Joey, Pete, and David. These are my ten men.

Are you pleased, Steve?"

"Sure, I am, Rugeey. Let's introduce them to Robert, but be aware that we have Alckamus, one of the aliens, trapped in our room back at the hotel. Don't worry. He is contained, which means he can't escape."

"Well, if you say so, Steve, I trust you." Five minutes later, they arrived inside the lobby of the hotel, in front of the elevator.

"Rugeey, do the honor of pressing the fourth floor button. We're going to room 420."

They arrived at room

420. "We are here, guys." Knock, knock.

"Hey, Steve, who are these guys? These are the ten most dangerous guys who are skilled fighters. The best in the world. We have Alex the Great, a skilled fighter in any style of fighting. Our other fighters are

Amber, Michael, Joseph, Jayden, Eric, Louis, Joey, Pete, and David. They also possess other fighting styles similar to Alex the Great."

"Wow! I'm pleased. You guys outdid yourselves. I'm all done with the weapons. I also have protective shields."

"That's great, Robert. I hope the weapons actually work. Did you test them already?"

"Yes, I have tested already."

"Okay, great! Now the ten men have to train."

"Rugeey, yes. They do have to train, Steve."

At that time, some noise came from outside, accompanied by bright lights. "Rugeey, stand back. I'm going to look out the window."

"No, Robert, don't do it."

Robert, Steven, and Alckamus all vanished suddenly.

“Hey, Rugeey.”

“What, Jayden?”

“Look at the sky and read the message.” It stated, “If you want to see your friends again, come and get them. Alckamus.”

Rugeey decided to create a strategic plan to rescue Robert and Steven from the spaceship that was once known as Mount Olympus.

"Guys, step inside. I've got an idea on how we can rescue Rob and Steve. We need to rescue them before the aliens decide to kill them and then us. Okay, this is how we're gonna do this.

Jayden, David, and Michael, you guys are gonna go in using the undetectable suits armed with the locator guns and grenades."

"We will wait for you guys here. If any problems arise, use the button that Robert added before he was taken. It's the teleporter button on the suit. It will bring you guys back to us. Guys, let me know if you guys are comfortable with this assignment."

"Rugeey, on behalf of all, we will die with honor serving you, Rugeey."

"Thanks, Jayden. Well,

now that we have that straightened, guys, good luck, and make us proud. We depend on you to bring them back safely."

The men went in the spaceship undetected by the aliens and were walking through the ship.

"Michael, what are you

looking at?"

"I see them. Rob and Steve are in that room, right next to Alckamus room. Be very careful. Let's try and find the keys from his room while he is sleeping. Okay?"

"I'm going to do it."

"Are you sure, David?"

“One hundred percent okay.”

David quietly pushed the button to remove the electrical field that was protecting Alckamus.

“I got the keys to open the door for the holding cell that contains Rob and Steve.”

David asked Rob and Steve, "Oh, are you guys okay?"

"Well, we feel funny."

"Don't worry, guys. We will get you out of here."

Steve said, "Hold on. These bars are made of electricity. As long as you flip the switch from the red box downward, the power shuts off. Go ahead, Jayden. Do the honor."

"Okay, Steve, my pleasure."

"Very good, Jayden. Now, Michael, you take Steve. I will go with Jayden. Press the teleporter button and bring us back to the hotel, okay? Let's go."

Everybody returned back safely with no physical injuries. "Well, it's good to see you again."

"Rugeey, that's why you we're chosen to lead this operation. Great job on getting Steve and me back."

"Any time, Rob, you

would do it for me."

"Yes, I would."

"So, guys, what happened up in the spaceship? You guys seem a little tired, and you have been acting a little strange."

"Well, Rugeey, Alckamus instructed his lead scientist Venum to inject a small chip in my finger. They said, 'It won't hurt a bit. Please cooperate and fall into the system. It's more painful if you don't fall into the system. This will help you see much clear through our sight.'

Meanwhile, back at the ship, Alckamus asked,

"Venum, you idiot, how could you let them escape?"

"Well, before they escaped, I implanted in them with a tracking chip with audio and visual surveillance."

"Well done, Venum, thank you."

"You're welcome, Alckamus. Now all I

have to do is keep track of their every move."

"Naboko will monitor Robert and Steven. So they won't try anything reckless and cause us to kill them. Ha-ha!"

Meanwhile, back at the hotel room, Steven asked, "Robert, do you still have the pen device?"

"As a matter of fact, I do. Go get it, Steven. It's in the left top draw in the other room. It's about time we take back control of this nasty situation, Steven. Push the pen inward, and once it lights up with the word scan, let it go. The pen will scan the room, including us, for any other device and deactivate it."

"No! Alckamus, hurry at once. You must see this screen. The humans have deactivated their chip."

"Robert, that pen you created did the trick. So now, what do we do?

"We prepare for the worst, Steve. Just expect the unexpected. For all we know, the aliens are planning a full-on attack on us. We need to be ready at all times."

"You're right, Robert. Hey, Rob, do you still have that cloning machine?"

"Hey, Steve, get everybody right now. We're all getting cloned at least three times, okay? Now let's get that cloning phone machine. It's in the future. Let's travel back and get it. Alex, Rugeey, Robert, and I will be getting that machine."

"We're here at Robert's lab."

"Get the phone machine and get the hell out of here."

"We found the phone device now. It's heavy for a small object."

"Rugeey, it's made of fibro active matter, which is one of the most heaviest compound of matter."

"How's the phone work?"

"Well, Alex, you simply hit the green button as if you were making a phone call. Then place the phone on the floor and let the electric door appear and walk through the door. Let it scan your body. Walk out of the door, and depending on what number you pressed on the phone, that many number of clones come out after you step out of the door within a two-minute period. So if there isn't any more question, let's please leave and go back to the past."

Three minutes later, they arrived back to the hotel room 420.

"Now everybody, get together. We're making some clones. Who's first?"

"I'll go."

"Okay, Alex, go ahead. Press three on the cell and then hit the green button, okay? Now step out. Let's wait. Oh, there. They're coming out. Three of you. Okay, who's next? Um, Michael, go ahead. Two minutes later, three of you. Great! Who's next? How about, Jayden. Okay, after Jayden, it will be David. Next Rugeey and then Steven, Joey, Eric,

Joseph, Pete, Amber, and me. Okay, next, we will name them with the first letter of our names followed by a number. So, for example, Alex's first clone will be A One. Get it? So it causes less confusion."

"Hey, Rob! Steven! I got an idea. What if we

go undetected to the spaceship and kidnap Venum and put a chip in him to control him just like the aliens tried to do to you guys, and then we clone him and send the clone to the ship in place of Venum, and we'll have total control of him. So what do you think, Rob? Steve?"

"I think that's a brilliant idea, Alex the Great."

"Thanks, Rob."

"You're welcome, Alex. Well, Rugeey, assemble your team to the ship undetectable and get Venum and bring him back to me in one piece. He has to be able to walk through the phone clone machine."

"I know, Rob, got it. Going in for the task of getting Venum will be Alex, Joey, Eric, Pete, Amber, and I."

"Okay, Rugeey, good luck. Go get the undetectable suits and get Venum."

Five minutes later, Rugeey said, “Okay, be quite. While we’re going in the spaceship, the first two grab him, hit your teleporter button, and report back to Rob. Okay, let’s go find him.”

“We got him.”

“Great job, Alex and Joe. Now take him

back while I'll snoop around with the rest of us."

"Oh, look it's Naboko. Hey, do aliens sleep because he looks asleep"?

"Good question, Eric. Yeah, they do, but it's different. How they do it there is just weird. I guess that's why they're called aliens."

"Okay, grab him, guys. Now take him. Bye. We will catch up with you."

"Okay, Rugeey, you're the boss."

"That's right, Alex, and don't forget it. Come on, guys, let's go check out Naboko and bring him back too."

"Rugeey, do you think that's wise? We already have Venum."

"Yes, do it."

"But what about what Robert said. He said to get only Venum."

"We'll bring in Naboko. It will give us a big advantage against them to have control on the inside."

"Okay, I trust you, Rugeey."

"Thanks, Alex, and besides that, Rob and I discussed this last night as a bit of surprise to everyone. If we get Naboko, the scale tips in our favor. That's exactly what we want for our guys. Are you with us because if you're not, you're against us?"

"We are with you, Rugeey. You have our 100 percent loyalty, and I'll die feeling honored, fighting this war as you being our commander. Hey, Rugeey, weren't you the commanding officer for the army?"

"Well, as a matter of fact, I was the commanding officer for strategic planning. Oh wow, that's amazing. I never knew that. You never asked, Alex."

"I believe Naboko is in this room, Rugeey. What's the next move?"

“We grab him, take him back to our room to get the chip implanted in him, and then get him back to the same place we found him. Teleport us now. Hey, Robert, where do I put Naboko?

“Right next to Venum. I already implanted

Venum and deactivated all devices and chips that could track the aliens to us having him. Now it's Naboko's turn to be deactivated and implanted with my chip. I'm almost done with the scan. Scan complete. Now, Steven, implant his chip now. Before we clone them, I'm gonna take a look at Venum's memory bank and view his memories. We need to know all the secrets of our world. We need to know what happened to certain places in history, starting with this computer-run city of Atlantis on Venum's memory bank."

Computer responds, “Records show that there was a huge war between the aliens ancestors and the Atlantians, and due to the big event, the city suffered astronomical and catastrophic events simultaneously that caused Atlantis to sink.”

"Computer, where is Atlantis currently located on earth?"

"Under the Bermuda Triangle. Details of the event during the huge war, catastrophic and atmospheric events took place. There were huge earthquakes that erupted volcanoes oozing lava, tornadoes, and tsunami. All took place on the day of the war. Unfortunately, nobody won. The casualties were of a great number. Computer details of Mount Olympus records show that it's a huge, round cylinder with a bright, shiny light."

"Hey, Robert, that's exactly how their ship is."

"Computer details of inside Mount Olympus is very dark with ten rooms protected with electrical bars."

"Robert, did you hear that? We're dealing with possible Greek Gods."

"I think you might be right, Jayden. One more thing. We must ask for the memory bank. Computer, who are the actual Greek Gods Zeus, Haydees, and Posidan?"

"Records show that Zeus's real name is actually Alckamus whose power comes from the sun, and it allows him to create his famous thunderbolts.

Haydees's real name is Venum. His power comes from death and fear of humans.

Posidan's real name is Naboko. His power comes from water and any type of weather condition."

"Okay, Rob, what's next now that the chip is implanted?"

"Well, Jayden, we clone them and put the originals where they were found. Naboko and Venum will be back into their spaceship without a trace." Quickly, let's clone them now. Let's make Venum first. Dial one on the keypad and press the green button to activate the machine."

"Okay, Rob."

"Steve, while he is walking through the phone clone machine, get Naboko ready."

"Okay, perfect. He is now ready."

"Now send in Naboko. Perfect. Now Jayden and Alex, go in the ship and leave. Now leave them exactly where you found them. First Naboko. When Jayden and Alex get back, Rugeey and Steven, drop off Venum."

Ten minutes later,

Jayden and Alex returned. “Now it is time for Alex, Rugeey and Steven to drop off Venum.”

“Come, let’s go, Rugeey.”

“Okay, Steve, you got it.”

Meanwhile, back at the ship, Rugeey said, “Okay, be very quiet,

Steve. Place him quietly on the chair with his head facing down."

"Come on, Rugeey. Time to go."

"We held up on our bargain. I wonder if I could take on Alckamus."

"Hey, Rugeey, you're talking crazy. What are

you doing? Please do not take off your suit. Please don't do it."

"I have to, Steve." Rugeey started running until she had a visual on Alckamus. Alckamus had a visual on Rugeey.

"You dare come into my ship and dare challenge me."

"Yes, I dare challenge you, Alckamus."

"Challenge accepted."

"Jayden kept his distance and watched the fight between Alckamus and Rugeey. The two threw punches back and forth at each other. Alckamus decided to finish the fight. He charged up his body and sent a thunderbolt to the chest of Rugeey.

“Rugeey, please don’t die. Teleport me back to the hotel.”

They arrived back to their room. “What happened, Jayden?”

“She tried to fight Alckamus by herself. I tried to warn her. She refused to listen to me."

"Oh my! Please help her. The only thing I could do is take her brain and implant it in one of her clones. Do it, guys. Please save her."

"So now we wait. Hey, let's all sit down and have a reflection moment," Steven said. Michael said, "Well, I would like to tell everybody about myself. I remember when I was in high school, nobody even wanted to sit next to me. I liked this one girl named Emily. She paid me no mind at all, but I still felt compelled to talk to her. Until one day I said the hell with her. I don't need her. I could do better, so that's when I met Alex the Great. He was so bad, as in high school, he had all the girls lined up waiting to bang him. I asked him, 'Dude, what are you doing to get all these girls lined up to bang you?' He said, 'I eat some good cheese.' I said, 'Wow, that's it?' He said, 'No, I also have a big nacho. That's how funny he was back in high school.""

"Anything else you would like to add, Michael?"

"No, that's it, Steve."

"Who wants to go next? Amber, how about you?"

"Well, all I want to say is that I have grown from the last five years. I'm now twenty-five and have control of my own life. I remember the day I met Eric. We were in college and both clubbing at Karma's basement. We were having fun. When some random guy pushed me, I fell back and hit my head on the floor. I had heels on. I also fractured my ankle, but Eric saw what happened and rushed over to me and asked me if he could do anything to help me. I said no. Then he turned around and walked over to the guy and asked him to apologize to me, which he refused. He asked the guy one more time to apologize. The guy said, 'I ain't apologizing and what are you gonna do about it, fagot? The guy was rather larger in size than Eric. I knew Eric could kick his ass and send him to the emergency room. The guy swung at Eric. Eric ducked and came up with a nice open left hook to the guy's temple and laid the guy out. I just remember that day as if it happened yesterday. That's the day Eric and

I became best friends."

"Alex, would you like to share a story about yourself from your past now?"

"Yes, I would. When I was eighteen years old, my father was a four-time light heavyweight boxing champion. A four-time mixed martial artist heavyweight champion taught me everything I knew before he passed away. By the time I turned twenty-six, I already knew everything he knew. His last words to me before he passed was, 'Don't forget me and live life with no regrets. Fight for a cause and not for the hell of it. Appreciate everything you have around you because once you die, everything is gone forever.'"

"That's a very good and emotional story. Thanks for sharing that with us, Alex. I know your lives are on in your mummeries.

Eric, would you be kind as to grace us with some knowledge about yourself?"

"Sure, why not? Well, where do I start? Um, let me see. Oh, I have a great a story from my past. As a young child, I realized I was gifted. I had the strength of ten men and could run very fast, but I always had the urge to help someone in need. There is one more gift. I have, but I will reveal it to everyone when the time comes, but for now, this is all I have to share."

"Hey, everybody, I would like to share something about the

first time I met Rugeey. I hope she makes it from the procedure. Well, I miss her already. But when I met Rugeey, she was in the gym, working out, and I noticed she had a six-pack. I asked her, 'How did you get those six-pack?'

She told me, 'I work

out my stomach seven days a week for one thousand crunches a day.'

I told her, 'Damn, you have a great body. Could you train me, please, and how much do I owe you?'

She replied, 'What's your name?'

I said, 'My name is

Amber.'

She replied, 'Okay, I will train you for free, but we must become BFFs.'

I agreed, and that's about seven years ago. Wow! Time flies. All I know is that she has been good to me, and I don't wanna lose a great friend. Please don't let me see Alckamus in the flesh. I will rip his head off his body. I hate him. He took out my friend."

"Amber, she will be fine. Robert and Steven have been working on her for about three hours now, and I'm sure they're gonna do whatever it takes to keep her alive. So don't you worry your pretty little soul. Have faith."

"I do, Alex, but you just don't know how it is when something precious has been taken from you."

"Yes, I do know, and

at least, in your case, there's a chance that they could be brought back. I wish we had technology we have now years ago. Be blessed, very blessed."

"I'm sorry, Alex. I totally forgot about your father. We're all upset at what happened to Rugeey and forgot what you went through. Please forgive me."

"Apology accepted. Don't worry about it. Just trying to snap you back into reality. That's all."

"Hey, Joey, we haven't heard a word out of you. Your turn."

"Hey, I'm just a

Brooklyn kid. What could you expect? All I do is party work and bullshit a little. I used to play football as a kid. I was a left tackle. I was pretty good until I caught a knee injury in a championship game. The doctors said it was a torn meniscus. I refused surgery, but I did bounce back by going to all my physical therapy sessions. Thank god if he or she is out there because if it wasn't for him or her, I would've been cut open, and who knows how I could've ended up?

Pete said, "I would like to add to Joey's story. Hey, didn't I play you the finals for the championship?

“I believe you did. You’re the one who injured me.”

“Hey, look that was a long, long time ago. I’m sorry. We were both young, ignorant kids from Brooklyn.”

“Well, thanks for the apology, Pete.”

“You’re welcome, Joey. I would like to share my story.”

“Go ahead, Pete.”

“Well, I also lived in Brooklyn in the Park Slope area, but one thing that was my passion was wrestling and MMA. I learned both the techniques as a child from Alex the

Great's father. He never told me that he taught somebody else. He told me it was just me. I'm the only child he was training. He lied. Did he also tell you that I'm your brother?"

"No, he never mentioned that. How could he do that if

we're brothers? What is my last name?"

"That's easy. Agustine. Oh my! I can't believe this. So, Pete, that makes you my elder brother."

"That's correct, Brother."

"But it's not so bad. I guess I could learn to deal with it."

"Hey, Alex, you have no choice. I'm your brother whether you like it or not. So I guess, we both are gonna have to deal with it whether we like it or not. Do you agree with what I just told you, Brother, or should we not speak to each other and be savages toward one another?"

"Listen, Pete, if you're my brother, that means you're my blood, and blood is thicker than water, which means your blood that passes through your body is the same that passes through mine. So I never turned my back on family. I'm not about to start now."

"If it's okay with everybody, I would like to share some personal information with everyone about myself. Hell, if I were going to try to stop aliens in a battle, I would like people to know who they're fighting side by side with."

"Go ahead, Jayden."

"Okay, well, I have supernatural abilities. As a child, I was exposed to high toxic chemicals which, at first, made me sick, but eventually, I had powers you wouldn't dare imagine, like, I can fly, I can't be killed, and I can burn people with my hands. I also can disappear and reappear whenever I choose to. It can be a good thing, but it also could be a bad thing. As I have these powers, people treat me very differently, and it hurts really bad."

"Oh honey, come here, Jayden. I think you're awesome, and you're okay in my book. If anyone thinks you're different, they can sit on a tack or choke on a bone."

"I hate when people talk a lot of junk for no reason at all. It makes me sick to my stomach."

"But, Jayden, just remember one thing. You have a gift. Not many people possess the abilities that you have. I wish I could fly and burn people with my hands. There's a couple of people I wouldn't mind burning, but that's for another time and day to be discussed. But for now, just be cool, and don't let people get the best of you. With great power comes great responsibility. Remember that. Don't ever forgot this discussion, and if you do forget, stop and think. Try to remember this day, and you look back on this day and laugh. I know you will."

"You're right, Amber. I'm a fool for even letting people persuade my judgment into thinking that because I'm not like them, I can't be part of their circle. Maybe there's a good reason why I'm not in that circle. There's a reason for everything."

"Hey, David, what about you? Anything you would like to share with us?"

"My gifts are simple. I can get in your head, find out what you're thinking, and persuade you to think a different way on any topic. I mean I believe it's a good gift. It's as great as Alex's or Eric's, but I have some tricks up my sleeve, and as a kid, I practiced and toned in on my skills. By the time I turned thirty, I perfected and controlled my gifts."

“Thank you, David. That story is appreciated greatly.”

“You’re welcome, Steven.”

“Now, Joseph, it’s your turn, buddy. Would you like to share anything?”

“Yes, I would. I remember going to school being ten years

old and being bullied all the way to school. I was scared back then. One day, I decided to fight back, and boy, did I let loose! It was so bad. I kicked three boys' asses at the same time. It was not for the weak at heart. I gave one kid a black eye, the other, a broken nose, and the last kid, I left him unconscious. They all had what was coming to them for messing with me. They had no business to mess with me, but I stood up to my bullies the way Rugeey stood up to Alckamus. If we wanted, we could take them. We're evenly matched now. Are you guys with me?"

"Not yet, Joseph. Trust me, Alckamus's days are numbered."

"You really think so, Steve?"

"I guarantee it, brother. Okay, now let me go and check in on the Rugeey update. I will be back, guys."

"Hey, Robert, what's going on with Rugeey?

Is she going to make it?"

"Well, as we speak, she is not going to make it, and we are gonna have to bury her. We'll put her in an electrical grave containment area. Go ahead, Steve. Grab her head. Let's place her on the table and press the button now. (A button was pressed.) I can't believe she is dead."

"Yeah, but she won't be forgotten. Steve, quickly check the cameras from the chips that are in Naboko and Venum. Let's see what's going on in the ship."

"Well, it seems that

Venum is talking to Alckamus about leaving earth and the humans alone. Alckamus is enraged and wants to hurt Venum now. Naboko is stepping in to calm the situation and gets zapped by Alckamus. Alckamus tells Naboko, 'Let me handle Venum the way

I see fit, and don't ever get involved like that again.' Alckamus notifies Venum that he must send a hostility gas over room 420 and that it has to be a strong enough dose. He wants to see the humans against each other. Robert and I are trying to come up with a plan to divert the gas back to the ship while making believe the gas has affected us. I got it. Robert, why don't we make a gas repellent barrier that would repel the gas back to the ship instead of hitting us? What do you think of that, Robert?"

"Well, don't get me

wrong. I like your idea, but how are we going to make a gas repellent without arousing suspicion of our knowledge regarding their plans? Wait. Hold that thought, Steve."

"We will make them invisible and small and undetectable so that they won't know what hit them."

"So would you like to start on the project, Steve?"

"How about tonight. They'll be ready by tomorrow."

"Sounds like a great idea, Steve. You know, Steve, you're the kind of guy everybody wants to be around.

You're cool, calm, collective, and can handle stress way better than I can. That's something I always admired about you."

"Well, thanks, Robert. Let's make it into a grenade."

"Okay, you got it."

Ten minutes later,

Robert said, "It's done."

"Lay it in front of the house. Set the grenade for protection of the house. Okay, press the button, and let the countdown begin. Ten nine eight seven six five four three two one. Look at the bright lights flying over the house. So now we're protected. So if these damn aliens try anything, we're ready for them. I like being a step ahead of them.

Meanwhile back at the spaceship, Venum asked, "Alckamus, are you ready to drop the hostility gas among the room 420?

"Yes, Venum, do it now."

"Okay."

Ten seconds, nine, eight, seven, six, five, four, three, two, one. Smoke was sprayed and repelled.

"Hey, it is not working."

"You imbecile, you do

not know anything, Venum. Why did I even choose you?

“Why are you talking to me like that? Have you forgotten all the battles I saved you from dying?

“I wish and recommend that you change the way you speak to me. I will not take any crap from you, Venum."

"Who do you think you are?"

"I'm Alckamus, the general, the greatest commander of this ship. Do I have to take you out like I did that human girl? Well, answer me. Do I?"

"I'm not going out

that easy. I can sense your fear, Alckamus. I'm getting stronger. I can feel your power entering my body. How do you feel, Brother? Are you fully weak yet?"

"Stop, Venum, please stop."

"No, you said you're the general. Show me

something."

"Hey, I'm getting weak now. What are you doing?"

"Alckamus, I'm not too fearful now, ha-ha."

"Hey, Robert, look at the screen. You have Alckamus and Venum going against each other. We can't let them kill each other just yet. Send them a frequency that will calm them down to the point they fall asleep."

"Ready when you are, Rob."

"Okay, Steve, send the frequency wave."

"Waves sent, Robert."

“Okay, back to the screen. It should take effect immediately.”

“Oh, I see it. It’s working now.”

“Wow! That was a close one. Venum and Alckamus are now both sound asleep. Hey, Steve, just monitor their brain patterns, will ya? I’d appreciate it."

"Okay, I'm on it."

"While you take care of that, I have to get everyone together and find out who's gonna be the new general for our war? I see two possible candidates, Eric and Alex the Great. Everybody come to the room. We need to discuss an important matter as we know our former general in war and combat has passed away. Rugeey may she rest in peace now. Well, we need someone to replace her. Someone brave enough to try and fill her shoes. I've chosen two candidates, Eric and Alex the Great. Now choose wisely. All votes and decisions are final, so everyone cast your votes now. I'm passing around a paper and a pen. You choose whom you want to lead us into victory, sweet victory. I see everybody has pretty much made up their minds. So I will read the votes and come back with the winner's name. Okay, guys, here it goes. The new general to lead us into victory will be Alex the Great."

"Hey, Alex, you might have won the votes, but in my eyes, you're no leader, please. I deserved to be the general. You call yourself the great. Show us how great you are, Alex I challenge you now to a fight. Come on, fag. Let's do it, punk. You don't have the balls. Very cool. Ya'll just choose a pussy to lead your squad. Good luck with that. I'm outta here. I'm going somewhere useful where I'm appreciated."

"Hey, I'll go with you."

"Thanks, Amber. Anybody else coming with me?"

"I'm coming too."

"Michael, how could you leave us?"

"I thought Eric would've made a better general. Sorry, Alex."

"Oh no! Hey, we gotta let Eric be the leader."

"It's too late, son. All decisions are final. I said that earlier. Well, Alex, you can't. Please, everyone, let's get some rest. We have a big day tomorrow and that's reorganizing our priorities and goals."

Eric, Michael, and Amber were walking toward the spaceship. As they approached, they encountered Alckamus standing in front of the ship as it hovered over them.

"Alckamus, we come

to join and serve your cause. That's destroying Alex the Great and his team of idiots."

"Well, humans, I see your logic. Go bring me ten gold items right away, and then we will discuss your agreement as far as Alex is concerned."

The three went in search of gold items. They all remembered where they could find gold.

Eric arrived at room 420 along with Amber and Michael. They all were looking for the secret safe that Alex had hidden, which contained five gold watches and three gold rings.

“Eric, over here. I found his safe.”

“Thanks, Amber for finding now. Michael, open it now.”

“Okay, Eric, you got it.

Pow! A loud explosion came from the other room and woke up

Alex.

"Hey, what are you doing here, Eric?"

"I'm taking your valuables and bringing them to Alckamus."

"Hey, you can't be serious you're working for him. You're so pitiful, Eric."

"Then stop me, Alex.

I'm right here."

"No problem. Once I beat you, you have to open the safe door. Now that Michael died, I don't need his death to go in vain."

"Agreed. If I win, you have to let me hit you with the relocator gun and send you to Mars, and if I lose, not only will I open the safe, but I will open the safe in front of Alckamus while the ship hovers over my head."

Eric charged at Alex, picked him up, and slammed him to the ground. They're both fighting. While on the ground, Eric caught Alex with a strong hook to Alex's temple and knocked him out.

After a long fight, Alex lost to Eric.

"Okay, you win, Eric. Take me with the safe to Alckamus. I'm going to open the safe."

Twenty minutes later, Eric brought Alex to Alckamus with the safe. Alex opened the safe.

“Okay, Alckamus.”

Alex opened the safe and gave the safe to Alckamus.

“What is this? It’s not gold. It’s silver. Damn it.”

Alckamus powered up and struck everyone

including Alex. Everyone die. Back at room 420, the real Alex woke up and found one of his clones was missing. So he went through room 420's hidden cameras and found that his clone A1 was fighting Eric and losing and was taken away along with his safe with no gold.

Alex told everyone about what happened regarding Eric and his clone A1.

Robert said, “I think I know where they might have gone. I think they went back to the ship. Let’s hack the ship’s camera network to see what happened."

"Hey, Robert, did you hack the cameras yet?"

"Yes, in one minute, we will be able to see everything that went on while we were asleep." There was a moment of silence.

"Oh wow! That's crazy. They all died."

"I'm afraid so, Alex. Your clone A1 has been eliminated. We have two left. A2 and A3 are the only remaining clones left. Be mindful. The phone clone machine as well as the time machine have been destroyed by Eric when he invaded the room."

"Wow, Robert, so what does this mean?"

"Well, Alex, it means we have to beat Alckamus. There is no turning back. When we beat Alckamus, we will automatically get placed in an alternate reality, which will give us a fair chance of evolving the way. We should've evolved on our own terms."

"Wow, Robert, we really need to stop Alckamus because if not, we will be stuck here forever."

"I understand, Alex, that you have deep concerned for being stuck in the past, but at least, the human race will be controlled by the human race, not the alien race. We're giving ourselves a fresh start."

"So I get what you're saying, Robert. If we kill Alckamus, we most likely kill ourselves in the process, but we will have an alternate life that includes a life without aliens."

Meanwhile, back at the ship, Alckamus said, "Venum."

"Yes, Alckamus, I just killed those three homosapiens. Get rid of their bodies. Pour some decomposition liquid to get rid of the body remains."

"Okay, I'm on it."

As Venum dissolved the body, he got an idea to invade room 420 and kill Robert and Steven.

"Alckamus, I have an idea."

"What is it?"

"What if we invade 420 and kill Robert and Steven and take their precious weapons?"

"Sounds like a terrible idea. Do you know the amount of security that is protecting that house?"

"The only way we can infiltrate that room is by sending robotic soldiers in and capturing Robert and Steven. That's the only way that's gonna work."

"Let's do it, Venum. Send the robotic soldiers now to capture them. Go at once."

The robotic soldiers arrived at the hotel and, in one second, appeared in room 420 and set off the alarms.

Ring! The whole team awakened.

“Oh my goodness! What is this, Alex? Look it’s a bunch of robots, and they’re wrecking everything. We need to stop them get them, Alex.”

“Already on it, Steve.”

Alex went Berserk at the sight of seeing

those ugly robots.

Alex threw uppercuts, jabs, left hooks, right hooks, and the special move. He called the one, two ankle breakers and caught the special robot by the name of Kunos right on the jawline and knocked him out cold.

Afterward Jayden got cornered by two robots, and Alex stepped over to rescue Jayden from the evil. That was the robot clan of soldiers. Now Jayden and Alex kicked some robots' ass along with Joey and Pete.

They're were now

getting the best of the robot army and were schooling them in the process. During the battle, two of the robots snatched Robert and Steven and took them to the ship. It was all a distraction to get Rob and Steven. Back at the ship, Rob and Steven were in the lab with Venum.

"So Robert and Steven, you two cost us a lot of grief and headaches. You're gonna die now. All you have to do is drink this liquid, and you'll die."

"Are you pleased, Alckamus?"

"Yes, I am."

Steven and Robert both drank the liquid

substance.

"Oh my god! I'm in so much pain, Robert."

"Me too, Steven. Aw! I'm changing."

"Me too."

Steven and Robert had both mutated into aliens.

"Alex! They're missing."

“Who’s missing, Jayden?”

“Robert and Steven. They’re gone. The aliens have captured them. We need to come up with something to save them.”

Okay, the whole team was coming up with a solid plan to save their friends Robert and Steve.

“I got an idea, Joey. Watch the perimeter of the ship and report any findings of what you see. Take this two-way radios, and keep me updated. Jayden, watch the back of the ship, and keep eyes on any strange movement going on. A2, stand about forty feet away from the ship and make a lot of noise to get the attention of Alckamus while Jayden's clones enter and search the ship for Robert and Steven. Everybody, take your positions now."

Back at the ship,

Alckamus was ordering the robots in prisons to escort Robert and Steven to the electrical jail cell.

“Venum! What is all the trickery? For you never told me that you made a substance that turned Robert and Steven into what we are. You have to die now."

Electric bolt electrocuted Venum, killing him at once.

"Now tell me, Robert or Steven, how does it feel to lose your friends to my thunderbolt and be back in prisoned again? You're not gonna beat me ever. I have an army of soldiers. The leader of my soldiers happens to be an alien hybrid named Kunos. I believe you have met him. We haven't lost a war yet, and as long as he's our general, we won't lose ever, especially to our creations ha-ha!"

“Well, Alckamus, we will say this. We won’t back down even if you created us. Prepare for a long and tedious road because we will not stand down, no matter how everything is stacked against us. We will prevail, and be ready for surprises of your life. We have tricks up our sleeves.

Hope you can handle them."

"You damn humans, don't ever speak to me that way again."

"Or what?"

(Alckamus began to charge his body when he heard a loud noise coming from outside of the ship and started walking toward the ship.

“Naboko, go out there, and check back with me. Let me know what’s goin’ on.”

“Yes, Alckamus. I’m on it. I will send the Kanuts to attack them.

(Naboko sent an alien-type dog to attack A2 and the rest of the team.

Jayden told A2 to run and get the hell out of there.

As A2 ran, the Kanuts started chasing A2. There were about five Kanuts chasing him. As that happened, Jayden sent J3 to confront Naboko.

J3 was also accompanied by E1,

L1, L2, M1, and A1 and were asking Naboko to step down as he walked toward them. Naboko told Alckamus, “Sorry, but my true allegiance is with the humans. I never liked the wars that were waged against them. I always thought they’re not fair. They never had a fair chance since we tampered with their DNA."

Alckamus showed signs of weakness and needed gold to survive.

"I'm getting gold now."

"No, Alckamus, do not turn on the gold magnetic attraction."

"What is that, Naboko?"

"If turned on, all that is gold will be attracted to this ship. Oh no, there is a gold shop, not too far from here. Damn, this is not good."

The ship attracted a lot of gold, including bricks, watches, rings, and necklaces.

"This is not good if he eats all that gold. Besides the sun, he will be way more three hundred more times powerful."

"Thanks, Naboko."

"Anytime, Jayden."

"So what else can happen, Naboko?"

"He can summon other aliens from our planet through his electric bolt. At his best, he could do huge damage."

"Well, Naboko, do you know how to stop it?"

"Well, I have to think a moment."

"Okay, but make it fast. I'm going to let

Alex know what's goin' on. Alex, pick up. Go for Alex. Alex, it's not good. Alckamus has gold and is starting to eat it, and if that happens, Naboko said that he can summon other aliens from his and other planets here. Basically they can be transported through his electric bolt."

"Jayden, thanks for the update. I'm coming over there now and bring the gold desintegrator grenades to blow up the gold. We will divert him till I get there. We don't want him to eat all the gold. I'll be there in five minutes."

Alex got there with the

rest of the team and clones.

"Rugeey 1, 2, and 3, get his attention. Joey grab the gold. Pete distract Kunos, and if anyone needs me, I'll be here watching, and if anyone needs help, I'll send reinforcements."

Everybody received

their orders. All went to perform their tasks.

“D1, D2, and David, Joseph, J1, J2, and all come with me inside to the electrical jail cell to hopefully find Robert and Steven.”

Alex’s plan had worked. Alckamus had been distracted enough where he was not powerful enough to bring more aliens from his planet to earth, and he and his team were inside the ship and arrived at cell to see two aliens.

"Alex, oh, are we happy to see you?"

"Who are you, alien?"

"I'm Steven, and he's Robert."

"No, that can't be. Okay, you're Steve. Tell me something only Steve and I know."

"Okay, I remember Alex telling me how before his father passed, he had trained him with everything he knew."

"Oh no! What have

they done to you guys?"

"They made us drink this liquid substance that turned us into what they are. We can also shape-shift. Wanna see?"

"No, not right now, guys. We have to get you outta here. Hold on! Let me turn off the electrical switch.

Alex turned off the electrical switch.

The robot soldiers came and saw them trying to rescue Robert and Steven. Quickly, Alex used the relocator gun now.

“I’m shooting them. They’re all going to Mars. Ha-ha!”

All sent most of them to Mars, frozen.

"I'm glad we made out safely. Come let's take you back to the room 420."

As they arrived back at room 420, Robert and Steven got an idea.

"Hey, guys, group meeting. How about we fix our laboratory

better than before and make stronger weapons than before we need to. Alckamus is just too strong. Even when he's weak, he is strong. I will create everybody a suit, not only that makes you just as strong, but that will keep us protected from his lightning bolt, giving you the ability to shape-shift and take their powers and making them yours. What do you think about that, guys?"

Everyone said, "Hell, yeah," with cheers of joy.

"I have a name for that suit. I will call it the all in one laughing out loud.

Steven and Robert worked on the suits for days, and it took two weeks for completion of the suits.

"Group meeting, guys. Today, we will be field-testing our suits. So, everyone, try on your suits and grab a partner. Okay, Alex will practice on M1. Get ready go.”

Alex shape-shifted into Michael just by simply touching him on his left shoulder.

“Fantastic! It was a success. We need one more test. I simulated a robot that has similar capabilities that

Alckamus has like his signature lightning bolt. So Alex, our fearless leader, go touch him as he's charging up. Then we will know if this suit truly works. You will possess his powers."

The robot began to charge up as Alckamus would. Alex walked

toward him and touched him. Alex said he felt the electricity in his body.

Alex hit the robot with an electric bolt. "Okay, here it goes."

Alex hit the robot with the electric bolt and fried his circuit breaker.

"Alex, great job. The

suit works now. We have to find out how the suit will stand up to a real battle. Hey, everybody, Eric, what are you doing here?

"Well, remember? I had a special gift."

Everyone was in shock on seeing Eric walk in with Michael and Amber.

"Well, the secret is I can't die. I'm immortal."

"Oh my! That's crazy on all levels. Tell me about it."

"I also have the power to bring people back to life. That's why Michael and Amber are both here with me as well."

“Alex, could you please come in here? I have something I need to share with you.” Alex stepped into the room.

“What is it, Eric?”

“I apologize for the way I treated you. I should’ve never doubted you as leader. I’ve been seeing you handle this team in a way I never saw coming from you. I underestimated you, and I'm willing to accept you as my leader. Please forgive me."

"Apology accepted."

Now that Eric was back on the team, their chances of winning this war have just improved by a large margin. Alckamus, better be ready for what's coming.

"Alckamus, we have to decide what will be done with those damn humans."

"Well, Kunos, we need to find out their weakness."

"Well, Alckamus, I am two steps ahead of you. We cloned Eric and controlled his memory, and he is now working for us on getting us all the humans' weakness and strengths. So we can build on that."

I am pleased with you, Kunos. You have

pleased me. I never knew you possessed the intelligence you have shown me today. Leave my sight and keep me posted on Alex and his team."

"Yes, at once, sir. I'm on it."

Back at room 420, everyone's joyful and pleased to have Eric,

Michael, and Amber back, but Alex saw something very odd about Eric. He didn't seem hungry and told Alex that he hadn't eaten in days.

"So tell me, Eric, why aren't you hungry?"

"I just don't have the urge to eat."

"I want you to eat this

slice of pizza right now."

"Please don't. I'm not hungry."

"Eat it."

"No!"

"Alex, what's going on here?"

"Well, I can't prove what's wrong with him, but I know

something's wrong."

"Hey, Robert, Steven, when you guys turned into the aliens, do you still have an appetite to eat?"

"No, not regular food, but we could eat some right about now."

"Okay, I'm going to test this. Somebody pass me my gold

watch."

"Okay, here you go, Alex."

"Okay, now is this delicious? I bet you want to eat this gold watch.

"Um, so tasty."

"Come on, Eric, you know you want to eat it."

“No! Stop! I’m not hungry.”

Eric knew he had the urge to eat Alex’s gold watch, but he was fighting the temptation.

“What did they do to you?”

“To me? I . . . I don’t know. Is there something inside?”

"Let me know, Eric. We can help you. We're not your enemies. We're like family. Come let us help you. Tie him up and scan his body. We need to find if there's anything in his body."

Robert set the pen to scan Eric's whole body and detected that he had been cloned while deceased.

“Oh no! How do we deal with this? He has been cloned while being dead. Do you know, Robert, how serious this is? He will be highly unpredictable, and his dreams will always be about himself being killed by the lightning bolt."

"This really sucks, Alex. Oh, I have a solution. We can reprogram his brain so that he thinks it's a bad dream and doesn't realize he was ever dead."

"Good idea, but will it work, Rob?"

"Yes, it will, Joey."

Back at the ship, Alckamus was growing impatient as each day passed.

"Kunos, get in here now. I need to speak with you at once. I'm not pleased with you. As of late, you need to watch the screen from which Eric is recording, and if this goes wrong, you're dead meat."

"I promise that everything will go as planned. I will make sure of it, Alckamus."

"Oh, and Kunos, if I hear that you're trying to dethrone me as your leader, it will be off with your head."

“No, I would never do that. Why would you think that of me?”

“Just making sure. I heard some of the soldiers tell me that they shouldn’t listen to me because Kunos is going to take over when Alckamus dies. So for your sake, I hope that isn’t true.”

"Never, Alckamus, my allegiance is to you, only you."

As soon as Alckamus left, Kunos communicated with Robert and Steven and asked them if they could work together in killing and getting rid of Alckamus.

"Well, I don't know if

we could trust you, Kunos. I mean a few weeks ago, you were destroying basically everything in our room and trying to kill us."

"But I had to. It was the only way I could survive. Alckamus is becoming very paranoid, and it's making me very nervous. We need to get rid of him."

"Okay, but first, we need to know if you're telling the truth, and if you are, we'll work together. Okay, sit down in this chair. Put your finger in this device. Now every time you lie, it will show up on this screen, and at the same time, you'll be zapped with some electricity something your boss Alckamus is very fond of."

"Okay, let's begin. First question, do you want to work with Robert and me?"

"Yes, I do."

"Okay, he is telling the

truth on that one. Next question, does Alckamus feel intimidated by his creations?"

"Yes, he does."

"Another right answer. He's telling the truth. One more question. Did you clone Eric to eventually betray us?"

"Yes, but that was

Alckamus's idea."

"He is telling the truth."

"Okay, take everything off him. This is how it's gonna work. Kunos, act like you're still with him. We will take it from here on our part. Wear this chip and put it in your ear. Every time you tap it while it's in your ear, I will automatically know what you're thinking and saying, and it will come up on my screen. So I advise you not to screw us on this. We will know if you do."

"Okay, guys, thanks. I will not mess this up. I want to be in charge of planet Nubu, and this is my meal ticket."

"Wait! You want to take over Nubu. That's what this is about. Hey, you better not screw us on this because if you do, you will not hear the end of it, and we will come find and get you wherever you hide.

Trust us on that. Am I clear?"

"Yes, I understand, humans. As leader of planet Nubu, I will have no interest in earth or humans. You have my word."

"Oh, before you leave, Kunos, I want to know everything about Alckamus,

Venum, and Naboko."

"Well, Steven, here, take this chip. Twirl it around your fingers, and a hologram presentation will appear and explain everything to you. It interacts with you and answer you based on what you ask. I took the liberty of naming the system Nubu. The reason is we have all our planet's history on it. So you will learn a lot from the system and as well as how the planet and our resources function."

"Thank you so much, Kunos."

"I'm also on it. So enjoy yourselves."

"We will."

Ten minutes later, "Robert, let's check it out. I wanna know about Alckamus, Venum, Naboko, and the planet Nubu."

"Nubu, tell me everything in detailed information regarding Alckamus."

"Alckamus. He's

twenty feet tall and weighs about six hundred pounds, and all muscle he has is approximately sixty-five thousand years old in human earth years."

"Nubu, did Alckamus ever have a significant other?"

"Yes, her name was Dunia, also known as

Hara in Greek mythology. She passed away five thousand years ago. He has never been the same again. After her death, he changed and became more angry and never remained the same again."

"Nubu, tell us more about Venum."

"He was the brother of Alckamus and always felt overshadowed by his younger brother due to his brother being the stronger sibling. He was a great scientist and technology major and a great innovator, unmatched to this very day."

"Nubu, tells us about Naboko."

"He was an excellent security strategist and great combat specialist for our planet. He protected Alckamus from any possible danger.

Alckamus had a lot of enemies and was not very liked on the

planet. There were several attempts to assassinate him, but Naboko always kept himself safe and out of danger."

"Nubu, tell us about planet Nubu?"

"Planet Nubu is quite larger than earth. Actually three times the size of earth. Our technology is superior to yours. By far, everything is larger, especially our alien race. We surpass the human race by far."

"Nubu, tells about Kunos."

"Kunos is a half-alien, half-robot hybrid, the only one of his kind. He wasn't always a robot. He used to be a full alien until he started dying. Then he was substituted with robotic body parts and equipment. He was once a war general for the army of planet Nubu. He is also Alckamus's only first-born son. During the war of Mount Nubu, Kunos was sliced in half by an evil alien by the name Zenos, which caused him to be created half as a robot, Nubu, who is Zenos. Zenos is Alckamus's father and former leader of planet Nubu. After the loss in Mount Nubu, Alckamus won full control of the planet and caused the death of Zenos."

"Nubu, tells more about Dunia."

"Dunia was married to Alckamus. They were together for about ten thousand years until Zenos murdered her in front of Alckamus, laughing so sadistically while her body vanished into the sky."

"Nubu, how did the feud start?"

"The feud started when Zenos asked Alckamus if he had intentions of ruling planet Nubu. In response, he said, 'Yes, Dad, I do want to take over when you're ready, of course, to allow me to.' From that day on, Zenos never looked at his son the same way again. He then imprisoned his son and told him,' As long as I live, you will never rule this planet. You hear me? You will never be a king."

"Nubu, does your race introduce a new leader

or what you call king?

“Simply the current king chooses his successor, or if he dies, his eldest son takes over.”

“Nubu, do you have any information regarding planet Mars and their civilization?”

“Planet Mars had a living civilization ten

million years ago before the planet self-destructed itself."

"Nubu, whatever happened to their population?"

"Before the planet had become uninhabitable, the Martians assembled a spaceship and traveled to earth, beginning by passing through a black hole in space so they could live among your race without raising suspicion from your kind."

"Nubu, when did Mars self-destruct?"

Mars self-destructed about four million years ago."

"Nubu, turn off now.

Hey, Steve, you never explained how it felt while you were transforming into an alien."

"Okay, while I was transforming into an alien, my insides turned hard. My mouth dried up. I felt like someone was stretching me in two directions, and I did feel more powerful, and I spoke to Robert right after the transformation. I asked him how he felt. He told me it was the same way I've been feeling. Thanks, Alex, for asking. Hey, everybody, check out what's going on the screen. Everybody's houses are being searched for any unusual activities for the absolute reason that the United States has banned cloning from ever being committed or being thought of, and anyone in possession of their clones will be confiscated and terminated."

At the door, they heard knock, knock.

"Open up. We need to search your room. We have a warrant by the United States government."

"Open the door at once. Alex, open the door."

The United States government found the

clones and asked who were the clones. "Take the clones and put them on a bus back to Washington, and we will be never seen again.

"I can't believe they're gone. They took all our clones. Now we're just evenly matched against Alckamus.

Hey, I guess, we are just gonna have to do our best against him and his legion."

"Hey, what do we do with Eric since he has a frail look about him and he reeks of death all over him?"

"Well, Jayden, I was thinking of putting him, Amber, and

Michael in a frozen containment tube."

"So basically he'll be frozen."

That is correct, but I'm also going to send them to space so they can stay there forever."

Robert and Steven had Eric, Amber, and Michael. All reported to the lab located in

420 and asked if they could help out with some testing. They all agreed. So Alex, Jayden, Joey, and Pete all helped them get strapped down and placed them in three separate capsules as each was done. Being frozen, they were sent to space. “Wow! I can’t believe they’re

gone. I'm gonna miss them."

"Yeah, me too, Pete."

"Hey, everybody, come together. I got something that I've been wanting to share with you guys. I'm lonely and haven't seen my girlfriend. I haven't seen her in months. Could I go get her and bring her back here, please?"

Everybody thought it over and agreed to have Alex's girlfriend stay with them until the war was over.

"Alex, before you rudely leave, what is her name?"

"You really wanna know, Joey? I will tell

you. Her name is Jennifer Collins, and thanks again for allowing her to be with us."

Alex the Great stepped into the lab to speak to Robert and Steve about which way was the quickest way to time-travel.

"So, guys, how can I

get there and come back here in seconds?"

"Well, we created a device perfect for time-travel, and you can also see your destination via hologram and also set your location via hologram. It can also scan any type of book or document and project it into a video via hologram. Watch and hit the stopwatch feature twice, and think about where you wanna be. At the same time, the hologram pops up. You'll be there so fast. Do the same to get back. But you also have to clear the watch."

"Thanks."

Alex arrived back in the future in his hometown of Queens, New York, and bumped into his girlfriend.

"Where have you been, Alex? I was worried sick. You even had your mother looking for ya."

"Well, that is why I'm here. Do I really have to tell you the long story or the short version?"

"Give me the short version."

"Okay, here it goes. I love you and missed you dearly. I went back in time with the help of a team to try and save our race from being tampered with by alien DNA. So basically, I'm here to bring you back with me to keep you by my side, where I can watch you, and if the past is affected, I get to spend my last few days with the one person. I'd rather spend them with the person I love the most. Besides my family, you, babe.

"Aw, sweetie, you made my day. Sure, I'll go wherever you're going as long as I'm with you, babe."

Alex and Jennifer held hands. He erased his watch and started it up and ended up back in room 420 with Jennifer.

“Everybody, this is my fiancée and currently my girlfriend, Jennifer, but she will be my wife one day. That I promise.”

“Oh, stop, Alex. You’re making me blush, baby.”

“I like when you

blush."

"Wow! Alex mentioned you to me, but meeting you in person is better than being described."

"Hey, Pete, Joey. Lay off. She is mine and mine for a reason. I'll kick both of your asses. Try me."

"Hey, we don't have

problems. Alex, we are a team. Remember, don't forget that."

"Hey, guys, I won't forget, but just show me some respect. That's all I'm asking for here. Jennifer, how are our children? I miss my kids so much. Well, how are Alex Jr. and Stephanie? How tall are they now?"

"Well, Alex is about five feet tall, and Stephanie is about four feet with eight inches."

"Wow! Kids grow too fast, especially mines. Hey, Robert, I had a crazy dream, and Rugeey was in it. She warned me that I have to be careful and not to make the same mistake. She took a lightning bolt to the chest and said not to underestimate him and he is stronger than he looks. So what do you think, Robert?"

"Well, Alex, I think you should listen to Rugeey's message."

"She also stated in the

dream that planet Nubu's past aliens and also Martians were sending messages back and forth to each other. I then asked her to describe the physical appearance of the Martians. Rugeey replied by saying the Martians are four feet tall zero inches with an oval head shape with black eyeballs and only have four fingers on each hand and foot. Their skin color happens to be red."

There was a knock on the door.

"Who is it?"

"It's Allen from the United States secret service. We need to come in and speak to you, the occupier of this room."

"Okay, hold on. I'm opening the door."

The secret service walked into the room.

"Hi, everyone, my name is Allen Hancock. I work for the United States secret service. I'm here on business matter.

I've been receiving report stating suspicious activity coming from this room. So we are going to need Robert and Steven to come with us to our office located in Washington D.C."

"Sure, Mr. Hancock. We would love to go with you to your office and discuss this matter."

Robert and Steven arrived at the office via horse carriages and started to get off their carries.

"Wow! We're at the site of what will be the future White House location. Today's date is July 4, 1776. It's very exciting, Steven."

"I agree. That means the president is George Washington, and the vice president is Thomas Jefferson."

Allen escorted Robert and Steven to their office, and as soon as they walked in, they saw George Washington and

Thomas Jefferson. They were in shock and pure amazement that they're seeing two of the four founding fathers of this country, sitting down ready to hear Allen speak to Robert and Steven.

"Robert, please answer some of my questions. First, what is goin' on in your room 420? Explain as much as you can, including all types of noises and bright lights."

"Well, sir, our light bulb shines very bright. Sorry about that."

"What light bulb? What are you speaking of?

"I mean, candle, sir. I apologize. I meant to say candle."

"Why did you mention light bulb? What object are you speaking of."

"Where I come from, it means candle."

"Okay, how about explaining the loud noise coming from

your room 420."

"Well, sir, we usually play games and have fun. In the process, we make a lot of noise. We try to enjoy our life as much as we can."

"Okay, I guess that will be all. If any further questions may arise, I will get in contact with you."

"Okay, sure, Allen."

As Robert and Steven transported back to their room, Alckamus arrived at the office of the secret service.

"Alckamus, no. I'm going to kill you."

"George, we had an agreement that

humans would not ever time-travel or mess with any technology. Your race is our slaves, and they're starting to tamper with our present moment."

"Alckamus, we can help you. Please don't kill us. You leave me no choice."

Alckamus charged up his body to unleash a massive thunderbolt, which killed George Washington and Thomas Jefferson, and escaped before the thunderbolt was released.

Meanwhile, back at room 420, "Kunos, what are you doing

here?"

"Well, I'm here because Alckamus just killed George Washington, and it gave me an opportunity to hand you this manuscripts. I do warn you guys some of your relatives are in these manuscript."

“Steven, go get the time-travel watch to scan this book, so we could view on the hologram, okay?”

Book was being scanned by the watch. One hundred percent complete.

“Okay, guys, we’re done scanning the manuscript. Hit the

holo mode of the watch. We should see it in two seconds. Oh, there it is. Wow, amazing how it's decoded and transferred into video format. So I'm going to ask this program about Atlantis."

"Sure, Robert, go ahead."

"Atlantis, show me possible relatives of Atlantians. My name is Jayden."

"Jayden, you are a descendant of King Noble."

"Atlantis, who is King Noble?"

"King Noble was a great king of Atlantis, who won wars through his vast technological skill set with that he amassed a wide range of awards and achievements. As young boy, he was honored in death with a huge ceremony."

"Hey, Pete, come in here and ask Atlantis who were your descendants?"

"Atlantis, who was related to me? I'm Pete."

"Your descendant was Andrew, our best military and combat specialist and was King Noble's right hand. Even until death, he was at King Noble's side."

"Atlantis, it's Pete .

Whom am I related to?"

"You're related to King Titan, a man who possessed the power of speed."

"Atlantis, it's me, David. Whom am I related to?"

"You're related to Linus, king of the thunder. He could

control thunder with his sight."

"Atlantis, it's Alex. Whom am I related to?"

"You're related to King Noble, which means that you and Jayden are probably distant cousins or brothers. King Noble's powers were flight, strength of ten men, and ability to take someone's ability and power."

"Atlantis, it's me, Louis. Whom am I related to?"

"You're related to Chameleon and have the ability to change into anything or anyone."

"Atlantis, what happened during the war between our ancestors, the Atlantians and the aliens?"

"The Atlantians were evenly matched with the aliens, and the power struggle of the war was too great for the earth to withstand

and caused the earth to sink Atlantis."

"Atlantis, do you think that we have a chance of beating these aliens?"

"The battle won't come easy, but if you're well prepared, I see why not."

Meanwhile, back at room 420, Robert and

Steven decided to train David and Louis with everything they knew.

"Okay, David and Louis, you guys want to learn how to create technical equipment. I have no problem showing you. Just listen up and follow me by doing everything we do exactly the same way. Today, we're creating a portal from an ordinary compact disc. Okay, first, I'm gonna need some microchips to attach to the sides and corners of the discs. Then we juice it up with electricity. Then you create an on-and-off button and then you're done. We also have to test it, of course, after it's officially done. Okay, Louis, turn on the switch."

As Louis turned the switch on, he turned into dust, and David got sucked into the portal.

"Ha-ha! These morons thought they could

take our spots. It won't be that easy."

Robert and Steven just tricked Louis and David into killing themselves.

"Quick, Robert, let's get rid of their remains. We don't need Alex and the rest of the team to find out what we've done. They don't have to know as long as you keep your mouth shut. You hear me?".

"I hear you, Robert, but let's get rid of the remains."

Steven just poured acid on the ashes Alex just stepped into the lab.

"Hey, guys, I have a

question. What happened to Louis and David? Haven't seen them. They said they were gonna hang out here and learn a few things from you guys. I hope they're okay. They really look up to you, guys. Well if you see them, tell them I was looking for them."

"Okay, Alex, you got it."

(Knock knock)

"Who is it? It's me, Allen Hancock. I have an urgent message for Robert and Steven. George Washington has been murdered by Alckamus, and we need you guys to bring him back to life."

"Sure, I created a liquid substance that jump-starts the heart and gets the blood pumping back to heart, but before we go, I need Naboko to be freed from the holding area he is in, and we need to make a clone out of the president for his safety. We're gonna have to upgrade the hologram watch so it can create the ability to make clones for us at our disposal."

"Agreed, gentlemen. As you guys work on making the upgrade watch, I will look for Thomas Jefferson. He is in hiding."

Meanwhile, Allen Hancock is searching

for Thomas Jefferson and gets a lead on his whereabouts.

“Hey, Thomas, get out from behind that carriage, please. I will not let anyone harm you. We’re taking you back to room 420.”

“You told me I was safe before and look what happened.

George died. How can I be sure that you can ensure my safety?"

"Well, before I left 420, Steven gave me some weapons to help fight against Alckamus. So fear not. We have a fighting chance."

"Okay, I'm trusting you, Allen, with my life."

Allen arrived at room 420 with Thomas Jefferson.

“Okay, the upgraded watch is ready. We are gonna clone Mr. Thomas Jefferson for your safety.”

“I guess you’re right, but what is cloning, Robert?”

“Well, cloning is done

when a person or an animal is being copied through a device. So in other words, there will be another you."

"Oh, I see, but does the procedure hurt?"

"Mr. President, it won't hurt at all. You don't feel anything. Just relax and let the blue light move up, down, and around your body."

(ZZZZZZZZZZZZ!)

"All done. Wait a few minutes. The watch is completing its cycle. Now complete, sir. There you go, Mr. Jefferson. Your twin is here laughing out loud. I'm a funny guy, ain't I?"

"Right, Steven, yes, you are, Robert."

"Wow! Amazing! He looks just like me."

"Well, Mr. Jefferson, he is you."

They arrived at the secret service office via carriage.

"Oh here it is. Let me give him this. It will

bring him back to life."

George Washington woke up from death. They brought him back to room 420. They arrived there at this moment.

"Hey, let's clone George Washington at least once."

"Okay, Steve."

George walked through the machine. Two minutes later, his clone came out. The clones were taken back to the office as if nothing ever happened.

"Robert, we must keep the real George Washington with us."

There was a shadow

standing behind Thomas Jefferson. Steven and Robert walked toward the shadow.

“Who are you creature?”

“I’m Soku from planet Nubu, brother of Alckamus.”

“Everybody, get in position to attack him

but hold your fire. Now, creature, we are going to ask you only one time. Are you with your brother or against him?"

"I am not with or for my brother. I despise him."

"Explain, Soku."

"At about the time of our youth, our father

Zenos told us that I would be inheriting his powers. As the transition period reached our planet I would be the next in line to receive my father's powers, but somehow my brother became very envious of my progression and did everything in his will to take our father's power from me. This must have happened while I was in my resting period."

"Why did you travel back through time?"

"Well, Steven, if you must know, I usually travel through time. This was what I was left with unfortunately, but I cannot or have no impact on the past or present, but I can help you defeat Alckamus if I'm allowed to do so."

"Soku, welcome to the we-hate-Alckamus club."

"I'm glad to be of service for a great cause, and that is to defeat him. We are going to put a stop to his reign as ruler of planet Nubu and earth. We will no longer be his slaves."

"Naboko, how have you been, old friend? It's been a long time, friend. I see you couldn't deal with my brother's nonsense any more. You made a wise decision to leave. Do you remember when we voted against enslaving the human race and Alckamus tried to have us killed?"

"I remember, Soku. That was a very crazy moment in time, but I'm willing to risk my life for the humans so they could have a better one."

"I agree. I will do the same. It's for a better future and for our planet as well. Slavery is not the way to go and should never be in place."

Back at the spaceship, Alckamus got contacted by the

Martians, and they asked him, "Alckamus, do you need our assistance in battling these slaves?

"I would not mind. Just send them a laser attack to let them know how we take over in a battlefield situation. These humans have never been in an all-out war, but we have marten."

"You're right, Alckamus. Do you remember the war in Atlantis. That was one of earth's major wars, also known as the Atlantic war, and there were a lot of casualties. Among them were King Noble and King

Titan."

The Martians were en route to room 420 to laser bomb the building.

"Hey, Alex, Jayden, everybody, look at the window. There is bright light, and the floor is shaking. Oh no! What should we do? Is this an earthquake? Oh no!"

"I think we are under attack. Get all your gear. We have some aliens that we need to take out. Naboko and Soku, go and infiltrate the ship so we could take over that ship."

"Okay, we're on it."

Two minutes later, Naboko got past their

security team and was in the main perimeter of the ship)

"Okay, Naboko, let Robert and Steven know we are going to hack into their computer system due to its complexes and advanced security grid. Let's override its system."

"Okay, Soku."

About two minutes later, Naboko got into their system, which caused them to have control of all the ships including all its weapons.

"Hey, Steven, we have the control of the ship. So teleport on the ship now to take care of the other Martians. Tell Alex the Great to have some fun on kicking some Martian ass."

Alex the Great arrived at the ship and encountered a couple of the aliens and started to crack his knuckles. One of the Martians charged at Alex. Alex stepped to his side and tripped the four foot zero inches Martian knocking him down to the floor. Then Alex began to mount on top of the alien attacking him until he murdered him. He took out a few aliens.

Now Alex and the guys start to walk

around the ship. Now they took over and were happy that they could travel anywhere they pleased. Robert and Steven started to bring all their laboratory equipment to the ship, and they were gonna create the lab so that they would be comfortable there.

It had been a few days, and Alex and Jayden began to learn how to fly the ship.

“Alex, this is fun.”

“I know it is, Jayden. Let’s go to the lab and check in on Robert and Steven.”

Five minutes later, they arrived at the lab.

“Hey, Robert, how’s the teleporting device coming along? It’s been a few weeks since I last checked in with you. So let’s test it being that you have a better lab now with better technology.”

“Okay, Jayden, I’m ready to test it.”

As Robert turned on

the machine, the black hole appeared.

“Jayden, be very careful. It’s not wise to go in without the proper protective gear. Here put this on. I set the destination for Mars. Whatever you see, let me know. I gave a special two-way device that keeps you in touch with me while you're on Mars. I will be shutting off the black hole for a few minutes. I don't want to keep the portal open for a long period of time due to us attracting the wrong attention. Do you understand?"

"I do, Robert."

"I will be setting the timer to put back the portal on after ten minutes. So keep track of your time, but don't lose track of time, and don't go too far from the same location of the portal. If you lose site of the portal and if you do, you will be stuck on Mars forever. Please we need you to stay with us."

Robert turned on the machine the black hole portal.

Jayden stepped through the portal. "Wow, this is great. I see Mars. It's just as red as it is on the pictures. Wow, this really is the red planet. Martians, they also look red. What else do you see, Jayden?"

They contacted via two-way radio device.

"Even the dirt on the ground is red. I'm looking up, and the sky is also red, now I understand why this planet is called the red planet. It's amazing. Robert, pick up. Go for Robert. I really love this planet. It's a beauty. I'm going to mark the ground so I could look around."

"Jayden, I wouldn't do that if I was you. What if the aliens mess up what you marked on the ground?"

"I know, and I hear what you're saying, but

I want to explore this planet. We can do this another time."

"Okay, Robert, I understand."

"I'm going to send you the black portal as we speak."

"Okay, Robert, I'll be standing by to receive this portal. I haven't moved yet."

"Jayden, move a little bit. I don't want to lose in the black hole."

"Okay, Robert, I'll move."

"Jayden, you're a very important piece in this team."

"I'm glad that I'm seen in that manner."

The portal arrived

right on time.

"Hey, Jayden, welcome. How was Mars?"

"Being there up close it was a wonderful experience and I would love to go back there. All the aliens that live there are all small about four feet tall zero inches, and they all look red. I think there are more things to be seen on that planet."

"Maybe we'll do this another time, but for now, all I know is that our technology works, and I'm happy about that. I'm also working on other devices like hologram technologies and certain game rooms on this ship called the game deck."

"I agree with you, Robert, and I am glad that you are an expert at your craft of science and physics. Robert, I have a question. How have you been feeling lately?"

"Well, lately, I've been

feeling a little angry. I guess drinking that substance and unlocking my alien DNA is causing an imbalance in my body. I've been working on a liquid substance to control my body during my rage cycles. I'm almost done creating the liquid substance. It should be done by next week. I’m also working on the liquid substance that keeps you living longer just as long as the aliens live. I believe we are making giant strides to improving our quality of life.”

“Well if I might add something to what you’re saying, in the last month, you and Steven have done so much in regard to technology. Hell, in my eyes, you're the man. You brought George Washington back to life, cloned him, and put his clone back at his central office as if nothing ever happened. That is bad ass. Robert you have to give yourself some type of credit and stop doubting yourself. In my eyes, you're history in the making, man. Feel free to disagree with me at any time. I speak the truth."

"I feel honored, Jayden, and humbled that you feel this way."

"I appreciate your honesty. Growing up as a little kid, my grandfather would tell me. Hey, don't forget what I'm about to tell you, Rob. Honesty is the best policy."

"Hey, Jayden, I would like to show the cloaking feature I added for this ship.

We can become invisible at any time. I also included a maximum shield protection along with the cloak so we not only will be invisible but we will be protected as well. I'm so happy about this."

Meanwhile, back at Alckamus's ship,

"Damn those damn humans! I can't believe that they not only survived my attack but they took over the Martians' ship. I hate all those humans, starting with Robert, Steven, Alex, and Jayden. I hate them all. I wish they never existed. Kunos, get in here."

“Yes, Father.”

“Why are they not dead?”

“I don’t know, Father.”

“They need to die. I hate them so much.”

“Father, in due time, you’ll will have your chance, and when I have that punk Alex the Great within my grasp, I won't hesitate killing him. He is worse than scum. I don't even know how to describe him. He won't win. I will curse him. I will make sure everyone around him dies first, and then when I get to him, he will die, suffering a long and agonizing death. I can't wait. Ha-ha! Our faith is on a course that cannot be avoided. Alex, your blood is mine. I will feed it to my pets."

Alex and Jennifer walked toward the ship. They were brought up into the ship by a holographic light, and once in the ship, they noticed how it looked more comfortable than being in room 420, and now the whole team decided to make the ship into the main headquarters. Everybody had a room there.

“Hey, Robert, I like how everything is

organized here. You and Steven have done a wonderful job with this spaceship. This could be our home as well."

"Hey, Steve, how have you been? Haven't seen you in a while."

"I know I've been very busy, you know, like always, just running around, taking care of business."

"I hear ya, Steve. Naboko, is that you? Wow, where have you been?"

"Well, as you know, I was locked in a holding cell by the United States secret service division."

"I see. Well, when did

you come back to us?

“I see you’ve been back by today or earlier. I’ve been out of the loop. Excuse me, everyone, for playing catch-up.”

“Hey, Naboko, who’s the tall alien thing next you?

“Oh, excuse me, this is Soku, Alckamus’s

brother."

"Why is he here? He's gonna try to kill us."

"No, Alex."

"He's gonna help his brother. Trust me. I know how this works. We think he is good. For a while, he helps us and then next minute, we're chopped liver."

"I assure you, Alex, I'm a friend to your race, and I will do my best to stop my brother. I have no hidden agenda."

"I hope so for your sake. Hey, Robert, I enjoyed going to Mars. Is it possible that I could go to planet Nubu? I would like to see how it looks. I will be careful, trust me."

"Okay, tomorrow. I'm gonna get some rest."

"Sure, Robert. I suppose I should get some rest too. We will meet again. Tomorrow sounds like a plan."

As Jayden walked away, Robert was plotting something

very evil.

I'm going to kill Jayden. I'm gonna go to planet Nubu and set hidden bombs. I cannot stand Jayden.

Robert went to planet Nubu and set up hidden bombs that would go off when stepped on. When Jayden steps on this bomb, he will go. Bye bye, baby. Ha-ha! Okay, all done. Now let me get back to the ship.

It's now 2:00 p.m., and Jayden and the whole team started waking up and made a breakfast to die for. Jayden walked in and skipped breakfast, jumped right into the shower, dried up, then got dressed, and ran right for the laboratory to be sent to planet Nubu.

“Robert! Robert!”

“What?”

“I’m here to go to planet Nubu.”

“Are you ready?”

“I’m ready.”

"Okay, let's send you to planet Nubu. Put all your protective gear and take the two-way radio. Be very careful out there. It doesn't look safe."

"I agree, but I will be fine, Rob. I trust you with my life. Please don't say that."

Robert started on the

portal, while Jayden put on his suit.

"I'm ready, Robert. I'm gonna walk through the portal now."

"Wait, Jayden. I didn't send the location through, and it has not given me feedback yet."

"Okay, Robert, you're

the boss."

"Now you can go through. Just make sure you have the two-way radio device so we can communicate."

"Okay, Robert."

Jayden walked through the portal and ended up in planet Nubu.

"Go ahead, Jayden.

Walk around and tell me what you see. How does the planet look?"

"Well, everything looks larger than what I expected, and there are nothing but ships hovering over the ground about five inches off the ground I guess the ships are also their homes, and they also have a holographic staircase coming down from the ship to the ground. It's amazing, Robert. What a sight it is! Martians also live here. How amazing! This planet's atmosphere is made of breathable air that being carbon monoxide instead of oxygen."

Poof a big explosion! Jayden exploded.

"I'm glad he's dead now. I don't have to see his face anymore. Ha-ha!"

Joey, Alex, and Joseph walked in to find Robert laughing.

"Why are you laughing, Robert? Let us know so we could

laugh with you."

"Oh no, it's something you guys wouldn't understand. I was actually trying to find out an equation, the same equation that I was stuck on in college, but I remember how one of my classmates told me, 'You can't walk before you crawl.'

I thought that was funny, that's all."

"You're right, Robert, that wasn't funny. Well, if you need me, I will be in my quarters with Jennifer."

"Hey, Robert, I want to know what is going on in our time. Could you check?"

"Sure, Joseph, I created an inter-dimensional screen. So once I turn the screen on, a holographic image will be displayed, showing a menu. What would you like to watch, Joseph?"

"Well, I would like to watch the news."

"Well, I'll program it for six o' clock."

"Cool stuff, Robert."

"Thanks, Joseph. Okay, the news is about to come on the screen. So let's sit back and watch."

The news anchor talked about how NASA was creating civilization for humans to live on the moon and also on all other planets.

"Wow, so much has happened in our absence. I can't believe that the human race will be soon living on the moon, Mars, and who knows what other planets."

"That's something I

never dreamed about as a kid. Imagine telling your dad, 'Hey, I want to live on the moon or Mars, Jupiter, maybe Saturn. Our species have come a long way, and that makes me feel very impressed.'"

"I feel the same way, Joseph."

"Thanks, Alex."

"I think I'm going to the future to check out the settlements along with my future wife, Jennifer. I will be back."

Alex, along with Jennifer, walked into the laboratory and explained to Robert that he wanted to go to the future for a little bit just to experience how the settlement of humans on other planets were coming along.

"Alex, I think you be teleported to the moon, so by the time NASA gets there, Jennifer and you will be the first ones there.

I'm setting the coordinates on the portal to be on the moon here. Alex and Jenn, put on these sunglasses and black hand gloves."

"Why am I putting these on?"

"Basically, it contains substances that can keep you breathing on any planet and can turn any gas or harmful gas into oxygen."

"Thanks, Steven and Rob, you guys are truly the best."

Alex grabbed his fiancée, walked through the portal, and arrived as NASA arrived.

Alex began to keep his distance while making sure not to startle any of the astronauts. Jennifer watched as the NASA team started working. As soon as they got there, first, they started building the base command headquarters. They had about ten people with them. All made strides to a brighter future. Alex became overwhelmed with emotion and told Jennifer that as a kid, he wanted to live on the moon and remembered telling his dad this, and his dad told him, “Son, in due time, this will be a reality. This moment was a historic one for Alex. Jennifer pointed to an object that was so large and traveling so fast. It's a meteorite and headed directly toward Mars. As Alex picked up the radio to tell Steven that the meteor hit Mars and split it into two planets, Alex started getting enraged and screamed at Jennifer. "Jennifer, what the hell is going on? We need to get out of here. Steven, pick up. Go for Steven. Please teleport me back to the ship. This sucks. I'll explain to you what happened when I get to the ship."

Alex and Jennifer

explained what happened back on the moon about the meteor hitting Mars.

"Hey, that meteorite was three times the size of Mars. When it hit Mars, a huge light shined right before the planet split. I wonder if the civilizations are okay. I bet when there is something evolving every time and when something like that happens, evolution takes place. Wow, humans were living there as well. I sympathize with the many civilizations from Mars. There must be a lot of emotional issues that have to be dealt with.

“For example, some of the families are probably on the other side of Mars and won’t be able to see them cause the planet split into two Mars. It’s so sad, very sad. It was crazy. I wasn’t expecting this to happen.”

“I did. Alex.”

"Why, Robert, did you send me there if you knew it was coming toward in my direction?"

"I knew as a fact that it was heading directly at you and only toward Mars. I apologize if I upset you with my remarks. I didn't mean to upset you. I would never intentionally put you in harm's way. I care about you and your family to do such evil things toward you. Please forgive me, Alex."

"Apology accepted, but choose your words carefully."

"Alex, there is something you should

know. I've been having problems ever since I became an alien, and I do not know how to control my rage or aggression, but I'm very close to coming up with the right chemical substance that can control my chemical imbalance. So all the medicines I've been taking have been making me more aggressive and sick, but it's the only way I can rule out the pros from the cons. So I'm going to the lab to tweak the last substance I have that is closest to working. I will be back in minutes. Everyone, I will let you know how that works out."

Robert went into his laboratory and began working on his aggression medicine. Twenty minutes later, he yelled with joy.

"Everybody, come in here. I found the right way to complete the medicine that will work the right way.

Okay, guys, I'm taking it now."

Robert took the medicine and immediately calmed down.

"It is figured out. So this is big news. Okay, so from now on, all I have to do is keep up taking my medicine. So I'm gonna need everyone to remind me. If I don't get reminded and forget to take them, I could become very violent. I don't know what I'm capable of. So, everybody, please help me. Alex, I would like to run your blood through the computer's database to check your DNA structure out of curiosity. Please, Alex, it won't hurt at all. The procedure is quick and painless. Just a simple prick on the finger. Once I receive the blood samples, I will run your samples through the computer, and in minutes, you will know one hundred percent who you really are and what you're really made of."

(DNA analysis is completed.

"The results are back, Alex. You're fifty percent Atlantian descendant and fifty percent of alien descent."

"No! I can't be this. Something must be

wrong with your computer."

"No, it's accurate, one hundred percent accurate. Let me tell you, Alex, this is a Martian ship. They're technologically more advanced than us, so this must be hard for you to stomach at the moment."

"But it gets better with time to adjust. You don't know how it feels to know you're an alien I was just getting used to the fact my bloodline descendants are from Atlantis, but I'm also an alien."

"Well, Alex, I could sympathize with you.

Look at me. I'm now one hundred percent alien. I have to deal with this every day. I'm trying to calm my aggressive nature. It's very hard. You know why I chose your name to be casted for voting as our next leader? Because you possess a unique ability that I admired. When in the heat of danger, you stay calm. That really impresses me. So if I were you, I would take that with a grain of salt because I don't like to give people credit, and I just gave you some."

"Thanks, Robert. I guess you're right."

"I know I'm right,

Alex. Now bring in your girlfriend. I wanna check her DNA and make sure she is okay. Alex, if the results turn out to be that she, in fact, is an alien, please be there by her side to console her in her time of need. She is going to need you more than ever."

"I understand, Robert, and I will do my very best to be by her side. In her time of need, I will explain that it's not that bad because I'm an alien myself. It might just help her cope with the situation. It just might, Robert."

"So bring her in."

"Okay, will do, Robert."

Alex walked in to the laboratory with Jennifer.

"Have a seat, Jennifer. I would like to take some of your blood for samples to check your DNA."

"Sure, but could I ask for the reason?"

"Well, we just want to make sure you're okay. Alex just finished his checkup."

"Okay, Robert, go ahead."

Robert collected blood samples from Jennifer and walked over to the machine and placed the samples into the machine to be scanned and traced for its origins.

"Well, two more minutes until we find out the answers."

"What answers, Robert?"

"Tell her, Robert."

"Tell me what?"

"We are checking your DNA to trace if there

is any alien DNA inside of you."

"Wow, really? You should have asked me in the beginning?

"What do you mean, Jennifer?"

"Well, Alex, I'm sorry. I didn't tell you sooner. I'm an alien, actually, a Martian. Sorry, I never told you."

"Wow, all this time, at least, you knew you were."

"I didn't actually. I knew you were all this time. I scanned your DNA while you were asleep two years ago. Just didn't know how you would respond or take the news."

"I need to leave for a walk."

"Yes, come on, let's go."

Alex and Jennifer walked down the ship's holographic staircase and took a walk down and started walking near the 10 Beaver Street neighborhood.

"Well, now that they left, let me find out more about Martians' living quarters. Computer, give me a detailed description of the Martians' living quarters."

"The Martians live in their ships for many years. Their ships are their homes. Inside each home contains technology-filled laboratory, a few rooms, an area to clean themselves, and an area to dine."

"Computer, one more thing. Please give me a detailed description of the Kanuts and how they were created."

"Well, during the war

between Zenos and Kunos, Kunos had an alien type of dog called Dungeey which had very sharp teeth that can rip through metal. So during the fight, Zenos grabbed one of the Dungeeys and electrocuted it with his thunderbolt. After the death of his companion, Kunos became enraged, killing Zenos."

"Thank you, computer. You can convert into sleep mode. Thank you once again for your assistance." Sleep mode activated.

"Steven, could you stay in the laboratory for me while I get some rest? I need to charge my body?"

"Sure, go ahead. I have no problem looking over the laboratory in your absence."

Robert walked away into his room to charge. Soku came into the laboratory. Soku arrived and entered into the lab.

“Did you call me, Steven?”

“Yes, I created a new drink that enhances your body’s senses and gives you incredible strength.”

Soku drank the liquid. It’s the strongest acid you could find, and it’s burning through the inside of Soku’s . . .

Now that Soku is dead, one less threat I have to worry about. Nobody is gonna be able to stop us.

Robert walked in and saw a disfigured Soku.

"What have you done, Steven? You killed Soku for what reason have you done such a thing?"

“Well, I never trusted him. From the beginning, he has been secretive. Since he has been here, Alckamus has not yet attacked us. Think about that. Why is that he has not yet attacked us? What could he possibly be waiting for? An opportunity he knew his brother was here and didn't want to attack right away. Alckamus is very bright. Don't ever underestimate him."

"He won't attack his brother without a real cause, and you killed the only alien that could've bought us time and really prepare for this war. Thanks a lot, Steven."

"Damn, I never thought of it like that."

"Why did you kill, Steven? He was a friend to our cause."

"Naboko, it was a mistake. My rage issue from me turning into alien is out of control."

"I will give him the

same medicine I created for myself to Steven to control it."

"Wow, he was my best friend and a good brother to me."

"Steven, you need to get better and control the rage. Don't let the rage control you."

"I understand, Naboko. I deeply

apologize for taking the life of your brother, and I will do whatever it to make things right."

Naboko went to his room to rest.

I can't believe Soku was talking to me. He warned me that Robert and Steven are not to be trusted. Whom can

I trust? I think the only person I can trust has to be Alex. Naboko rushed over to speak to Alex.

"Alex, I need to speak to you at once."

"What is it, Naboko?"

"Steven has killed Soku, and Robert has been killing everyone on your team. Jayden is one of many on the list compiled by Robert. Be very careful. What should I do? I'm very frightened."

"Hold on! Let me think, Naboko. Just try to stay near at all times, okay?"

"Okay, Alex."

"I will protect you,

Naboko. Let me leave the ship for a little while."

As Robert walked down the ship, "Robert, I have been trying to reach you. The clones are doing a fantastic job as the fill-ins for George Washington and Thomas Jefferson.

Well, I'm afraid it's gonna have to be a permanent position because we need to send the real George and Thomas into the future for further protection. Do you understand me, Allen?"

"Well, I guess, you're right, Robert. You

guys are the best, and thanks for protecting our presidents."

Robert went back upstairs into the ship.

He bring the presidents into the laboratory. We're sending them to the future frozen, and they must remain frozen for eternity.

The presidents were placed in the holographic freeze tube and relocator containment center to be sent to their future with directions of not unfreezing them.

Robert turned the machine on. Please set the coordinates to

Washington D.C. secret service headquarters. They are going to be blown away. I know, but who cares? Coordinates are set, and they should be frozen in about two seconds and should be relocated now. Yes, we did it.

The former presidents

arrived in Washington D.C. and were greeted by a lot of secret service agents.

“Kevin, don’t get so close.”

“Wait, boss, I’m reading the note. It says, ‘Please don’t . . . ’” Kevin, the secret service agent, passed out on the floor from shock.

"Alckamus, get up. I need to speak to you."

"Yes, Brother, I am awake. How can I help?"

"I'm here to help you. I just visited Naboko. Now I'm going to speak to you."

Alckamus tried to hug

him and realized he was in spirit form, which meant he was dead.

"Brother, I feel deep regrets for what I have done. You should have these powers I possess. They rightfully belong to you. I had no business taking it away."

"Please, Alckamus, I moved on from that. I'm trying to give you one last piece of advice. Don't go to war against the humans. You will lose badly. Please do not go forward with the war."

"Spare me, Brother. You always loved these damn humans.

The smell of these creatures are horrendous. They smell like dogs. We created them. Technically we are considered gods to them."

"Please, Alckamus."

"I know you have passed away, and I know you're trying to

prevent me from glory, but I'm going through it and also very intelligent. Ha-ha!"

"Alckamus, even intelligent beings cease to exist."

"Alex, you're needed in the laboratory."

Alex just arrived inside the lab after Steven just called him.

"Yes, you called me, Steven? What do you need?"

"I would like to run tests on you to find out what is going to bring out your alien side?"

"Hey, Robert, where have you been? I've been taking care of some business, Alex.

Well, Alex, Steve and I want to know what makes you tick in a sense we don't want you to change into an alien of something that triggered causing you to go into a rampage state."

"I agree."

But they shortly found out that stirring Alex

emotions would turn him into an alien form with unbelievable powers. He will be more powerful than Alckamus, Robert, and Steven combined.

“I miss Soku. I know, Steve, you killed him.”

“I apologize, Alex, but it wasn’t my fault. I couldn’t control my rage. I'm taking medicine for my imbalance."

"Alex, what is going on? You're changing."

"I can't control my emotions on how I feel about Soku's death. Now I'm changing."

Alex started changing into his alien form, due

to the fact of his ability of not being able to cope with his emotions.

"Robert, I got it. Why don't we create a ring that he wears that can trigger and control his emotions, thus giving us the upper hand in the war?"

"That's a brilliant idea,

Steven."

While Steven and Robert created the perfect ring, Alex decided to get some rest. Everyone in his team was also resting and was having weird dreams.

"Oh, snap! That was a crazy dream. I can't believe it."

“You should believe it, Alex.”

“Who are you?”

“Funny, you should ask. I’m known by many names. The most common among your people is the dream master.”

“What do you want?”

“It’s not what I want.

It's who I need and I need you and your team to beat the crap out of Alckamus."

"Why do you want us to beat him?"

"He has been after me ever since I beat his brother for full power and control of Tarterus and the dream world. So ever since I beat Venum, he has been upset and wanted to overthrow me as the leader of the dream world and Tarterus."

(Tarterus is also known as hell.)

"Believe me, you do not want him in control. He will use those new powers for the sake of evil."

"So what will he use it for?"

"His plan was to keep the humans as slaves forever and mentally have control on them so that humans will never find the truth about their existence."

"What is our existence?"

"Well, Alex, your existence was to be slaves, nothing more, and you have been distorted from the truth."

"Wow, and even and down to your precious religions that were created by Alckamus. Ha-ha!"

"I find that he has

always been that way and will never change."

Dream master was twenty feet tall and was able to fly, and he had a lot of dark features. He was basically a flying, tall shadow.

"Dream master, are you leaving now?"

"Yes, I am."

"Where are you going now?"

"I am going to see Alckamus and scare him in his dream and give him a good scare that no one or being would like to see themselves lifeless. Remember that I'm going to give him an image he won't forget ever."

Dream master arrived at Alckamus's ship.

"Alckamus, wake up."

"Dream master, what are you doing here?"

"I'm trying to clean what your brother left a huge mess. I lied to the humans in order to do, so they think I'm your enemy."

"Keep it that way, dream master. They do not have to know anything. I could use the help. Thank you, dream master. So what was the mess left behind?"

"Soku tried to tell Alex the Great about his

real past of Atlantis before he died, and if he did tell him and he decides to really find out, we're in trouble, Alckamus. Alckamus, do you still have his brother and father?"

"Yes, I do. They have been frozen right after Atlantis sunk, and we could never find him."

"Alckamus, once the young man finds out his parents are here, he will become obsessed and will be more determined to come find and kill you. What shall we do?"

"Let's send a carbon monoxide gas attack and kill everyone on that ship."

Alckamus arrived and unleashed the carbon monoxide that entered into the ship and started killing everyone on that ship, except Alex, Jennifer, Robert, Steven, and Naboko. They all managed to escape.

“This sucks. Alckamus just wiped out all our

team. Now what do we do?"

"Well, guys, let me think of something."

"I believe there is something I'm supposed to find out by the Bermuda Triangle. Soku told me something about my real family."

"Well, Alex, your

father that passed away wasn't your real father. He was a stepfather, and your mom was your stepmother as well. Your mother passed away a long time ago in Atlantis. But your father went missing."

"How did you know all this? Soku told me

he tried to tell you, but he was ignored by you."

"Wow, I never knew that. I wished I had paid attention, but things happen for a reason. How long does it take to clear this ship of these toxic gases?"

"It takes about twenty minutes."

“Thanks, Steve. Once the ship is rid of all the gases, we’re going to the Bermuda Triangle.”

“Why the Bermuda Triangle?”

“That’s when we can scan the ocean of any lost artifacts with our hologram watch to view any recordings of my father and possible brother if they're still out there. It's up to me to save them."

Twenty minutes passed. The ship had cleaned itself out of the poisonous gas. They boarded the ship and were heading to the Bermuda Triangle.

They arrived at the

Bermuda, using the ship's speed accelerator.

"Alex, I think you should be the one since it's your family and all to set the watch to scan and play back what's scanned from Atlantis."

Alex set the watch to scan. It was now

scanning everywhere and beginning to input all the information to be played back.

"Alex, the watch is ready to play what was scanned. Everybody, stay still and watch. Do not interrupt, please."

"Okay, Steve."

"I'm pressing the

button."

The hologram showed a couple Alex the Great Sr. and his mother, Valencia the Great. Alex the Great was an alien from Mars who lived on Atlantis and met Valencia the Great. They had a child who was the elder brother of Alex Jr., and his name was Luke. He possessed the strength and power of Hercules and was from the same bloodline as Alex.

During the war between the Atlantians and the aliens, the home of Alex was invaded by Alckamus, and his mother was killed by Alckamus with a thunderous strike. His father and brother both were captured. Alex was very young and devastated.

"Alex, are you okay?"

"No, I'm not. I remember that day now. It was locked away. I was very appreciative of inheriting my mother's flying and freezing abilities, and my father left me his fire flame thrower ability, and I can control any sea I wish to."

Alex and his friends went back to 10 Beaver Street.

Steven notified Kunos

that he needed to free and clone his dad and brother, so it did not raise suspicion that they're missing. "Okay, Alex, I'm working on that immediately. Anything else before I leave?"

"No, that's all, Steven, and thanks, man. I really appreciate this help and support from you and all the rest of my guys."

Back at Alckamus's ship,

"Kunos, we need you to clone and then free Alex's father and brother."

"That's great, but with what machine?"

"I have to build it."

"Okay, so get working on it ASAP, please."

"Yes, I will."

Kunos got all his equipment and started working on the clone machine.

"I got an idea. Kanut, fetch me that round disc. I'm going to create a cloning device out of that disc. Yes, I'm a genius."

Kunos walked over to the freezer section of the ship and pulled out Luke and Alex Sr. from their fridge. He carried them over to his laboratory to unfreeze both of them.

"Now all I have to do

is add a few more pieces to the disc, the antenna, and then on the switch, and now I'm done. These guys are almost done unfreezing."

Unfreezing completed.

"Guys, I need you to wake up. Oh, wait. I almost forgot. I'm gonna get them up to speed on how to talk in this language. Guys, drink this for me."

Luke and Alex Sr. both drank the substance and immediately caught up to speed and understand and speak modern-day English.

"Okay, gentlemen, please stand up and stay still. I'm going to have this disc scan your body, at least one time for each of you."

"Okay, Kunos, we're ready."

The disc started performing the initial scanning process, starting with Alex and then next happened to be Luke.

"I'm glad this is over

now. We have two clones. Let me freeze them and put them back."

Before Kunos left, he told Alex and Luke that he would be placing the clones in the exact place that they were and that they're going back to be with Alex. They are very grateful to be helped.

"Kunos, where are you?"

Alckamus looked for his son, calling him frantically.

"Okay, it's time to set these clones to freeze."

Kunos walked by and heard his father calling

and looking for him, so he hid from his father's view and sneaked right by his father into the laboratory.

"Okay, I need to get you guys out of here. It's not safe right now. Alckamus is snooping around, and I can't let him know I'm trying to help you guys out, so I'm going to teleport you to their ship. Alex, pick up. Go for Alex. Stand by to receive two."

Alex stood by his teleporting machine and was awaiting his long-lost father and his elder brother, Luke, who were both taken from him when he was an infant.

“Father, Brother, I miss you. Oh my god! All this time, I was told to believe that my stepfather was my real father and that was not the case, and now I’m face to face with my real father. I missed you so much. My elder brother, our youths have been taken from us by a cruel evil alien named Alckamus. Mark my words, he will pay for what he did to you two. If it takes my very last breathe when he and I meet against each other in our battle, he will never ever forget me, even his death."

"We lost precious moments that could have been but is not because of that damn alien. He will pay, Son. I'm glad to see you. I was heartbroken when I was taken from you unwillingly. If I could change the past, I would. I could make that happen. Son, please, things happen for a reason. Just leave it alone. The most important thing is that we are here now, and I'm going to enjoy every moment I share here now with you."

"Okay, Dad. I understand what you're telling me, but it's not fair. That's all I'm trying to say."

"Remember, Son, never cry over spilled milk."

"Okay, Dad. Luke, my elder brother, wow, I miss you too. We never got the chance to grow up together. All games we could have played together as kids."

"But I will have to

agree with Dad. What's done is done, so let's not dwell on all the negative things that has happened in the past. Let's build on a positive future."

"But, Luke, in order for us to do that, I have to kill Alckamus and his entourage, except Kunos. He is the reason for us being here together right now. So I thank him so much, and I'm blessed to have him on our side."

"Yes, indeed, keep your friends close and your enemies closer. That is true, my friend."

"So, Father, I'm trying

to use my powers that I inherited from you. How do I control and use it well to bring out the fire flame out of your hand?"

"Think of a place that is very hot. An oven, for example. Now think you're there and aim your hand at a target. Great, now think that your hand is a gun and you're shooting bullets from your hand. Go try it."

"Wow, this is cool, Dad. I like this."

"Now do the same, but opposite for water, think of a cool place, and that's how your arm will become a hose releasing an unlimited amount supply of water."

"Thanks, Dad, for showing me how to control my powers. At least, I'm more knowledgeable now than when I was two months ago. You're a very wise man, Dad."

"I guess we're both alike in a sense."

"Dad, Luke, stay here. I'm gonna go to the laboratory to speak with Robert and Steven."

Alex arrived at the laboratory.

"Hey, Robert, what happened to the suits that you created for us?"

"Well, dream master

took them from us and hid them, but, Alex just know you don't need those suits. You are learning and perfecting your powers, and this will work in your favor."

"I know, it will. I was just asking, that's all. I was curious of what happened to them.

That's all."

"I need to get in contact with Kunos so he can inform me about the dream master and his ties with Alckamus. I want to know everything about them. How they met? How long they knew each other?"

Kunos arrived to the

Martian ship to speak with Alex in regard to dream master and anything he would like to add after he described everything on dream master.

"Well, where do I start? Oh, right here. Okay, Alckamus and dream master are long-time friends. They will die together. That's how strong their bond really is. It's a shame how Venum was screwed in the process for dream master to take full control of the underworld and the dream world. Basically, dream master asked to take over from Venum. With no questions asked,

Venum is no longer the ruler of the underworld nor the dream world. It is very sad how their brothers, you would think, should be closer, but Alckamus is a very cold being with no emotions and doesn't know how to feel upset or happy. The only emotion he shows is angry all the time."

"Well, I appreciate all the information you have given me. Now is there anything else you might like to add?"

"Of course, I would love to add something else dream master was talking to Alckamus two days ago about lying to humans and that he can't be saving Alckamus ass all the time. Alckamus wants him to stay portraying himself as a nice guy. To screw all the humans royally, so if I were you, I would think of a plan to start taking out the dream master and Alckamus team one by one."

"You know, Kunos, you're right. Now I see why you won mostly all your wars. It's for being very smart. So where should I start? Kunos, you know the ship better than anyone here. Okay, computer, access Alckamus's blueprint for his ship."

Access granted. "Thank you, computer. I see that the easiest way is to start with the Kanuts. Their weakness is the smell of oil. Motor oil, but you lure them in and then electrocute the bastards laughing out loud."

"Wow, Kunos, I see

you're enjoying yourself a little too much."

"I guess I am, but it's not very often. I get to help create a strategic war game plan and effective one at that. This is in my blood, Alex. Well, now battery juice. I'm half-robot, but you understand what I'm trying to say, right?"

"I do understand. In my eyes, you're a great war general, and you're very intelligent more home."

"Thanks, Alex. Now what's the next target? Dream master?" Staying awake and giving the best of yourself for this fight, he is the weaker being in our world than his own dimension.

"That's why it is better to have an advantage, and fighting him in his dimension is not fair."

"So why would I ever put myself in that type of predicament?"

"Well, I guess that's

why you're the general and you're on our side cause. If you were not, I know for sure, we would have had a lot of problems because, my friend, you would have given me some big problems."

"Well, I will take out the first line of protection, which is the Kanuts, and then move on to the second line of protection, which is the dream master. After I take him out, he is all by himself."

"I wouldn't be so sure. Never count out Alckamus. He always, nine out of ten times, has something up his sleeves. I have seen him in action in many battles. I counted him out. I thought he was going to be dead. He came right back and finished his opponent off. He is very skilled with his signature lightning bolt, so when facing Alckamus, my father, you have to be very cautious. He packs a serious blow, deadly blow. I apologize to Rugeey and many of your friends that fell to the hands of my father. It saddens me to have to speak to you on such a very sad note, and if your friends were here, they would want you to beat Alckamus badly. So are you ready to execute your plan? Each time we meet, we plan for the next attack. Is this agreed?"

"Agreed, Kunos. You made a lot of sense. My friends would want me to beat him so bad that he will lie lifeless."

"One interesting fact about the Kanuts is that they were made by

Venum and Naboko, so your best road to travel is channeling Naboko since he is still living."

"You know, that's a great idea, Kunos."

"I will leave you now. I know you have to get back to your ship."

Kunos left and walked in the laboratory and

asked Naboko to answer a couple of his questions.

"Hey, Brother, I need to talk to you. I have to ask you some questions."

"Not now, Luke. I have to speak with Naboko first. Once I'm done, I will speak to you. Is that fair?"

“Okay, no problem, Alex.”

“Now, Naboko, how do eliminate the Kanuts?”

“Well, there are several possible ways. The first is attacking their complex circuits, which is nearly impossible to do since Venum placed various security detailing into their security system. The next and only possible way I see you stopping the Kanuts has to be by freezing or burning the circuits, and by what I'm hearing, that shouldn't be a problem."

"Well, I haven't fully mastered my powers

yet."

"I suggest you do it. It will help you, especially if you're planning to take on Alckamus yourself. I mean your father is back. You saved him. I think he owes you that. Have him help you zone in on your gifts. At this point, you're our only help at killing Alckamus, and if it doesn't work, as far as your father goes, you could always ask Robert and Steven to create something a device to help you control and monitor your powers at all times."

"That sounds like a

great idea. Thanks, Naboko."

"Great idea for which one?"

"Both of them."

"Which one are you going to choose first?"

"Well I decided to practice with my father to learn more about my powers, and then

I'm going to have Robert or Steven build me a device to control and monitor it. I'm glad I spoke to you today. You're a smart alien. I see why I recruited you to our team."

Everybody decided to rest until the next

morning. Everybody was in the dining room of the ship and enjoying their breakfast as Naboko walked very fast and, in a loud tone, told Alex, “Alex, I have an idea. Why don’t I go back traveling through time and meet with Venum as we were creating the Kanuts?

But I wouldn't travel regularly. I would travel in a different form. I would travel back in time as a spirit-type form, enter my body, and use that body as a vessel to execute our plan, and that is to beat Alckamus by first taking away his first line of defense, which is the Kanuts. I'm going to ask Robert if he has the device or machine capable of creating me in such a form."

"Okay, it sounds like a good idea. I'm going to the laboratory to speak to Robert for the device."

Robert was busy

creating a new wave technology.

“Hey, Robert, sorry to bother, but could you help in creating a device that turns me into a spirit form to enter my body in the past so I could tell Venum how I could access the security system of the deadly

Kanuts?"

"Well, Naboko, I will have to create it for you. Give me about a few weeks to make it. I will create a necklace that fits correctly on your neck, and it will have a button on the side that will bring you back and forth. Make sure that you have the information."

"I will, Robert. Do you need me to help you create the necklace?"

"It's okay, Naboko. That's why I got Steve."

Naboko left the laboratory, and as he walked out, Steve came in.

"Hey, Steve, we need to create another device a necklace that sends Naboko in a spirit form to travel back into time and into his body at the time period to warn and advise Naboko to help us get the security override code for the Kanuts to help us eliminate the Kanuts."

Two weeks had passed, and Robert and Steven had created the necklace. Naboko walked into the laboratory and asked Robert, "Is the necklace done?"

"Well, Naboko, try this necklace on."

"Wow, thanks, Robert, it fits perfectly on my

neck. Now how does it work, Robert?"

"Well, just push the button once, and you'll be there. It has already been programed with time date and location, so everything has been done. Just remember all you have to do to come back once you have all the information is click it once more. It's only programed for twice. That's it."

"So what happens if I press it more than twice?"

"What will happen? Nothing, you'll stay stuck in the past, so be mindful that the button has to be pressed twice."

Naboko pressed the button once and ended up in his past body.

"Oh good! I'm here."

"What do you mean here, Naboko? You've always been here. You haven't left yet."

"Oh no, Venum, I was thinking out loud

about something. Sorry about that."

"It's quite okay, Naboko. So are you ready to help set the security code? You're gonna set up your code, and then I will set up mine."

"Venum, how about we set up the code together while I

watch?"

"But we agreed that's how it would be. Why the sudden change, Naboko? Is something wrong?"

"Yes, I'm from the future, and Alckamus will kill you, and I'm here because I need to know the override code to stop

Alckamus."

"Well, Naboko, I don't think that can be done."

"Sure, it can. Did I mention Alckamus kills you?"

"How do I die? How does Alckamus kill me?"

"With a thunderbolt

right to the chest is in your destiny. It can't be avoided, so in that case, come and watch the code."

Naboko saw the code and told his friend Venum bye, and it was good to see him one last time.

"Welcome back, Naboko. How's

Venum?"

"He is great. So, guys, how did he die?"

"Remember, he killed himself. As he killed himself, Alckamus wouldn't get the codes. Smart, very smart. He knew he was going to die anyway. Very smart Venum. Guys, listen up. Venum didn't kill himself before I left. Alckamus killed him with his thunderbolt. By killing himself, he altered the future events and a different outcome possibly in our favor."

"Wow I never knew that. Yes, you did, Alex. I told you a while ago. So I have the codes. Everybody, oh, great, give them to me. I will scan them with the watch, and then when it's done, I will give the watch to Alex."

The ship experienced an earthquake-type reaction and caused the ship to move uncontrollably. The team looked out the window. It's Alckamus and his ship.

"Alex, are you ready?"

"I'm ready as I will always be."

Both ships created a large square holographic platform, on which anyone could stand on.

"Alex, I challenge you, but first, you must pass two challenges. If you can get past them, you can face me for the rights of your precious planet."

"Okay, Alckamus, you're on."

Alckamus sent his alien attack dogs to

attack the Kanuts, but Alex pointed his trusty hologram watch directly at the Kanuts, overriding each and every one of the Kanuts.

"Alex, be careful with the dream master. He likes to scare and intimidate once you get past that you can win."

"Thanks, Naboko."

All the team started rooting Alex on in cheers, but he faced a tougher opponent, who was the dream master.

"Alex, are you ready for your second challenge?"

"Yes, bring it, Alckamus, I'm ready."

"Your next opponent is the dream master. I hope you're ready to come with me to the other world, my world, Tarterus. Ha-ha!"

"If I were you, I would stop laughing. The only thing that is going back to Tarterus is you when I kill and send you back there."

Dream master unleashed all his demonic soldiers on Alex. Alex beat every one of them. Next, he sent a two-headed, one-eyed giant that walked on four legs after Alex. Naboko jumped in to help Alex and got struck with Alckamus's lightning bolt and was staggered by him.

Naboko managed to shake the attack off and waited with Luke and Alex Sr. Alex found the monster's weakness in his legs. So Alex looked over at Robert. He threw over a device like a bracelet and told Alex to put it on; that was going to help control his powers.

Alex put it on and could now control his powers and turned into his alien form to fight this four-legged creature. He immediately went for the legs by creating a huge fire ball that burned off the legs of the beast. He then proceeded to burn the rest of the beast's flesh until it was no more.

He then got close in the range of the dream master and was ready to deal with the dream master. One on one,

Alckamus was getting very concerned as Alex was going through all his defensive system. As Alex was gaining momentum, he beat the dream master by freezing him and then shattering the ice into pieces.

"Now Alckamus, it's your turn."

"I'm ready to kill you, Alex. I should've done a long time ago, you pathetic piece of space scum."

Alckamus tried to hit Alex with a lightning bolt, but Alex avoided it. Alckamus got hold of Alex from behind; he had trouble getting out of the hold.

Alckamus slammed him on the ground of the platform.

Alckamus seemed to be getting the best of him, but Alex seemed not to be moving. Alex Sr. started distracting Alckamus while Luke was checking on him, he asked Alex if he was okay. Alex said,

"I'm so banged up. It's not even funny." Luke saw that Alex had been wounded and told Alex, "I have powers to the ability to heal. Let me heal, please, Alex."

"Okay, take care of me, Bro." Luke healed his brother. Alex got up. Naboko got next to Alex, handed him an electric machete, and asked Alex, "Are you ready, Alex?" Alex nodded in agreement.

Oh Alckamus!

Alckamus turned around and saw Naboko lift and throw Alex while he was in his human form toward Alckamus while Alex had an electric machete in his hand and chopped off Alckamus's head, leaving his body decapitated. He then picked up the head and showed it off to his teammates.

"Hey, Brother, we have to get rid of that head. Its body is

growing back."

"Agreed, Luke. Let's put him in a containment tube and let him float in space."

"Sounds like a plan, Alex."

Alex and his brother are on their ship and placed Alckamus in a containment tube and sent his head with the growing body flying through space.

Meanwhile, back on earth, "What do we do with this body? It's growing a head. Let's implant him with a chip in him and keep him. We could use him, you never know."

"Yeah, you're right, Steven."

Before they could leave, Jennifer decided to ask Alex a question.

"Alex! Where do we go from this? How will we adapt to this massive change?"

"I can't answer these questions right now, but I guess, we will take it one day at a time."

Meanwhile, everybody walked toward ship and were headed back to the laboratory.

"Where do we go from here, Robert?"

"Well, I was thinking, maybe, we could go to planet Nubu and explore the planet."

"Yeah, that would be a great idea. Steven, off

to planet Nubu."

"Hey, shouldn't we consult with the team?"

"No, they will just have to deal with it, and another thing, we do not need to take all this medication. It is making me sick. How about you?"

"How do you feel after

taking the medicine?"

"I agree with you, Robert. We need to stop taking this medicine. I don't think it's healthy for us to keep taking this medicine, but I disagree with you on one thing. We should tell the gang."

"Okay, do what you

must, Steven. I will tell the ship's navigation system to take us in planet Nubu's direction."

Steven went and notified Naboko, Alex Jennifer, Luke, and Alex Sr. to all meet up in the ship's lounge area. They all arrived at the lounge area.

“I have to let everyone know that we are in course to go to planet Nubu on an exploration mission for a quest for peace and possible alliance between their world and ours. Does anyone refuses this mission?”

There was an awkward moment of silence.

"Steven, as long as I live, I will never deny this team. I will step up to any challenge, any mission, that is required of me."

"Well, Alex, that is why you were picked. I'm pleased to hear your response, and I congratulate you on your leadership abilities that gave us the victory we needed over Alckamus and his strong hold on the human race."

Twenty minutes later, "Hey, everyone, I need you to arrive at the main navigation deck panel and look how fast we have come so close to planet Nubu."

Two minutes later, the team arrived on the panel deck.

“Wow! Robert, this planet looks so huge. Are we going to land?”

“Well, we have to ask permission to land, but before I do that, let’s get rid of Alckamus’s other body. Let’s teleport him off the ship."

"Great idea, Steven. Let's take care of this matter before Kunos could ever find out because we wouldn't want him to violate the peace treaty we worked so hard for."

Alckamus was placed directly in front of the teleporting device.

"We will flip the switch in five four three two one, and he is gone. Yes! Okay, Robert, call Kunos and let him know that we will be visiting him so we can be embraced with open arms."

"Hey, Kunos, we are exploring every galaxy, and we're wondering if we could drop by and hang out in your hometown of new city on planet Nubu."

"Yeah, you're welcomed to visit me and my city. After all, I'm in charge now, and you guys helped me get there. I owe you and your team, especially Alex the

Great a whole lot of gratitude for helping me defeat my horrible father."

"Well, Kunos, I'm glad things worked out for the best."

"I agree. Alex, is that you?"

"Yes, it is, friend."

"Listen up! Everyone,

I need to share with Alex the Great and his team what happened in the months following me taking over new city."

Kunos started to share his accounts from what he told his father for the last time to the months after he became the new ruler of new city.

"Well, after my father's head was chopped off by Alex the Great, I walked over to him and told him how it felt knowing that I would be the new ruler of new city. He spat on my face. That's when I told your team they could do whatever you wanted."

"Oh yeah, I remember you saying the last part about do whatever you want with him, but I never knew what was said prior. Well, now I know."

"That's right, Alex, but what did you do with his head?"

"Well, we put him in a

tube."

"No! Tell me you did not do that. If he lands here, the tube will break. Nothing can hold up in this atmosphere. Alex, explore the planet while I come up with a solution to this madness."

"Okay, Kunos, but if

you need me, I will be outside of the ship."

As Alex explored the planet, he was fascinated by the fact that he was able to breathe better.

"Wow! This planet has better air quality than my planet."

"That's because your body can adapt very

well to our planet's atmosphere."

"Who are you if you don't mind me asking?"

"Of course, I don't mind. My name is Sophocies, and I'm just an honest being, trying to take care of a large family. It's very hard for me to do that now that Alckamus has been murdered. Everyone knows that his son can’t run this city as good as his father did all these years. What is your name, human?”

“My name is Alex.”

“Are you the human that murdered Alckamus?”

"Now Sophocies, I can help you take care of your family. I get along with Kunos. I will talk to him and make sure you're taken care of, okay?"

"I am trusting you, Alex."

"You have my word, Sophocies."

Alex was back at

Kunos's home ship and mentioned Sophocies and that he needed assistance to survive.

"Have no worries. I will take good care of our friend Sophocies, but in the event that Alckamus tried to take over the planet, I need you to kill him this time for good. Can you handle that task for me, Alex?"

"I understand, Kunos, and I am up to the challenge of taking out Alckamus."

"Now, Alex, turn on the screen. I want to know what's going on in my city."

Alex turned on the

screen.

“Okay, I love this news reporter. In my eyes, she is the best at what she does. So let’s listen to what’s going on.”

“Today, the bodies of Alckamus have been found off lake Venum as he appeared to be moving very slowly and very weak.

“I can’t believe what I’m watching since I’m in power now. I’m making an executive decision to have my military officers seize him and place him in two different cells.”

“Just now we have received confirmation that King Kunos has

given the military a direct order to apprehend and place Alckamus in a cell until they can figure out what will be done with him. I hope for our sake Kunos has not made a grave mistake."

The military had arrived at the scene of

the crash where Alckamus landed. They seized both of his bodies and placed them in the hover vehicle and drove to the military holding cell facility.

Their soldiers arrived at the facility to drop off Alckamus.

"General Dex, what

cell shall we place Alckamus?"

"Put him in isolation confinement 1 and the other body in isolation confinement 2, but keep heavy surveillance on him at all times. We don't want Kunos to get upset. He might send an order to have all of us executed. So be very precise, and do not let Alckamus escape."

"Sure, General Dex. Thank you, Private Grunt."

General Dex arrived at Kunos's home ship and discussed possible solutions for Alckamus. "Kunos, what will come of Alckamus?"

"I have a plan. You have to kill Alckamus in his cell. Tonight carry out and accomplish that task without messing up."

"Sure, Kunos, you can trust me."

"Don't make me regret it."

"I won't. You can count on me, Kunos."

General Dex arrived back at the facility where Alckamus was.

"General Dex Private Grunt at your service, awaiting your next direct order, sir."

"My direct order as follows: go into the cell where both bodies of Alckamus are and eliminate them both."

"But, sir, he was our leader and possibly could still be our leader."

"Mr. Grunt, all we know is that he went missing for a long period of time. In that time period, his son and next in line,

Kunos, became our leader now. I suggest you do as you're told."

"General Dex, I have very bad feeling about this."

"Just shut up and do as you're told."

Private Grunt stepped toward the cell of Alckamus, but Alckamus was awake and sensed something bad was going to happen.

“Alckamus, sorry.”

(Private Grunt shot off an electric gun and hit the holding cell’s electrical system and caused both Alckamus to escape, losing his left arm and merging with his other body.

"Private Grunt, get out of there. He is going to kill you."

As Grunt was trying to run away, Alckamus hit Grunt in his back with a lightning bolt, killing him instantly.

As Grunt died, the other soldiers escaped, heading back to Kunos to explain the whole situation of how they tried to kill Alckamus and failed and that he killed Private Grunt. "Let me get into this empty hover vehicle and try to find my old friend Theos. If anyone can help and make me better, he is the right direction I must go in. Vehicle, take me to Theos, the scientist engineer."

All the hover vehicles on the planet could operate on autopilot as Alckamus arrived at Theos's lab, he observed Theos leaving the laboratory.

"Theos, now where do you think you're going?"

"Alckamus, what

happened to your arm? Are you okay?"

"It's a long story. Well, as you know, my son took over the planet. He betrayed me on earth by joining the human rebellion against us and caused me to lose against a fierce opponent, whom I underestimated, Alex the Great, so simply put, I was split into two bodies. One was thrown in space, the other in a tube. Somehow our bodies ended up back here on an isolated part of the planet, and the media I believe broadcasted the fact that I wasn't dead, which caused

Kunos to locate me and throw me in a holding cell from which I escaped through a big explosion that caused me to lose my arm and merge with my other body. Does that cover everything, Theos?"

"Yes, it does."

"So, Theos, can you

help to restore my arm and my power?"

"Well, Alckamus, to be honest, you won't be the same. I will make you stronger than ever before."

"Well, I guess I'm going to have to deal with being stronger than I was."

"Alckamus, we need

you to defeat your son. He has gone mad, and I don’t know where his mind is. He is drunk with power just like you were when you first controlled everything in new city.”

“Promise me one thing that you will destroy Kunos.”

“I will try my best.”

“Once I was defeated new city law states that if a ruler has been defeated by any other being less powerful than their eldest son receives all their father’s strength and power.”

“Alckamus, I’m going to make you more

stronger and powerful than your son."

"Thank you so much, Theos. I really appreciate this, and I won't forget what you've done for me in my time of need."

Meanwhile, back at the ship, Robert and Steven had stopped taking their medications that kept them controlled for a week, devising a plan to take over planet Nubu from Kunos. They would send Kunos at a desert blindfolded and let Alckamus know where he was, so Alckamus could take care of his revenge while they took over planet

Nubu. Robert and Steven knew that Alckamus had been seen at the laboratory with Theos. A military personnel informed them)

“Hey, Kunos, Robert and I have a huge surprise for you. We know that we never celebrated that fact that you're the new ruler of new city, but how about we give you a surprise?"

"Well, fellas, I like surprises, so what would I have to do?"

"Just wear the blindfold until you hear your name, okay?"

"Sounds good."

"Hey, guys, before you leave, could Kunos show me all the rooms?"

"Sure, go ahead. Kunos, come on."

"Alex, I want to show you the room my dad used while he tortured and cloned humans."

"Okay, no problem."

"But fellas between us, that was his favorite room. He never saw humans as equals simply because we altered your DNA and improved your race. He saw humans as his slaves, and that's why he is in the predicament he is now."

"Well, guys, we are here. This is the room."

"Wow! This is crazy, seeing all these dangling people from the ceiling of the ship with no strings. How's that possible?"

"Well, he gives these pills that keep them levitated."

"How do you get them down?"

"Simply snap your fingers. Wanna see? That's okay. I believe you. Well, I must get going. Don't wanna ruin my surprise."

Robert and Steven placed Kunos in a hover vehicle and told a military personnel to tip off Alckamus on their location of Kunos.

"I'm setting the location, Steven."

"Okay, Robert. So Kunos, how do you feel about receiving your surprise?"

"Well, to be honest, I'm getting a little anxious and

impatient."

"Think of it this way. Soon you'll know the feeling of a good surprise."

They now arrived at the location, and it's the same location where Alckamus landed.

"Okay, stay right here. Alckamus will be

coming soon, and Robert and Steven, make sure they hide and keep a distance to be able to watch Kunos confront his father."

Meanwhile, back at Theos's laboratory, a military personnel delivers a message to Alckamus to meet

Kunos at the same deserted location where he was found. Kunos would be blindfolded ready for his attack. The message claimed that it was delivered to him from his faithful supporter.

“Alckamus, before you go, please test your

new arm outside."

"Let's go, Theos."

"Alckamus, you have to aim and then charge your body. The electric bolt is ten times more powerful than before because now you don't need gold to charge up, and the sun automatically charges up for you, and your eye that was missing has been replaced with an infrared red plasma beam that can burn through anything. If both are combined together, it can blow things up making a catastrophic explosion of mass proportions."

"Thanks, Theos. I must go to the

location."

"Oh, Alckamus, be careful. I hope it's not a setup."

"I believe that. I will be fine as long as I keep a visual surveillance of the area. I will be fine."

"I know you will, Alckamus."

Alckamus got into his hover vehicle that was given to him by Theos and arrived at the location within minutes. Before he walked over to Kunos, he stepped on a chip with a hologram message. “It’s from Robert. The message is coming from his hologram image. It says, 'When you see Kunos, yell his name.'"

"Kunos!"

Kunos immediately knew who it was and began to panic, and at this point, he realized that he had been set up by Robert and Steven and took off his blindfold.

"Yes, Alckamus."

"Why did you betray me? If it was power you needed, I would have easily stepped down for my eldest son. There was no need for you to have done this type of betrayal? For all I have done is be kind to you and take good care of you in the absence of your mother."

“Well, Dad, I needed the power for myself. I wanted to earn it on my own without being handed it to me. if you’re going to kill me, I’m ready to accept my fate.”

“Well, Kunos, it will be my pleasure to kill you and take back my city that was wrongfully taken from me by you."

Alckamus blasted Kunos with a huge plasma blast, exploding him to pieces.

"Wow, this pains that I had to kill my own son. Why oh mighty Hesiod, I pray to you after I have slain my eldest son."

Alex the Great happened to see Alckamus upset and walked over to him.

"Alckamus, I'm sorry that you lost your son."

"I didn't lose him. I killed."

"Come on, Alckamus.

Come with me. I'm bringing you back to the ship."

Meanwhile, Robert and Steven went to the nearest media, having recorded everything that happened, to purposely intimidate the media into letting the whole city know that Kunos had been murdered by their hands and that they're in charge of new city. This caused an uproar and riots.

"Alckamus, I know we had our differences, but now all we have is each other. Robert and Steven are in charge. Now that you murdered Kunos, and this is causing a lot of problems. For one, humans were never looked upon as equals. Now let alone they're the rulers of an alien planet. Alckamus, I'm only going to ask this once. Can I trust you to watch my back while fighting the same enemies?"

"It seems, Alex, I don't have a choice. If I do not have your help, Robert and Steven will ruin my city into the ground, abusing what I worked so hard to create a nation that the people are free from hostility. Alex, what is the goal? How do we beat Robert and Steven?"

“Well, Alckamus, we have to recruit an army. How about it? Do you have anyone in mind to bring in our favor?”

“Well, I guess, I can start with general. I did a lot of favors for him. I guess he owes me in a sense.”

“Well, will he help as a

definite?"

"Yes, I need to know. That's the only shot we have of going against Robert and Steven. You must remember, Alex, I never will underestimate them the way I did with you. I hope you realize this."

"I do, Alckamus, but we need to be victorious. There ain't no other way, and there's no turning back."

"We have the best army at our side, but we need weapons. Do you happen to know a technology major or scientist that can create weapons?"

"Well, as a matter of fact, I have someone in mind, Alex. I will advise my long-time friend Theos to help us. He has always been there in my time of need. I am blessed to have him. I will pray to the almighty Helios, our god. Here on planet Nubu, he will give me the strength I require for this battle. This is my ritual, and this is how I won my past wars."

"Alckamus, who is Helios?"

"Legend says Helios lived among us as one of my people. No one knew how he looked exactly, but he did extraordinary things like flying and turning our bad apples into anything he pleased. He also helped to create our planet, which I'm grateful for."

"So let me get this straight. Just because he can do things, you can't makes him a god."

"Well, I see why you would think that way? Well, how do the prayer rituals go and do you have a place of worship?"

"Actually, we do. It's called the house of Helios, where all are welcomed. He died for our sins."

"Wow, that is so similar to our Lord. His son came to earth and died as a sacrifice for the human race could go to heaven."

"That's just like the story of Helios."

"Wow, could this mean one thing, Alckamus? We share a

same god."

"You're right, Alex, and no matter what race or species, we answer to one god. That is so amazing."

"Alckamus, answer this question for me. How can you live in our conditions if your body was not created to live on this planet?"

"Well, I implanted my body with a body regulation chip on the inside of my brain, and it keeps my body's temperature at 128 degrees Fahrenheit. So I could be able to survive if I didn't do that simple procedure. I wouldn't be able to survive your planet's atmosphere. I also thank all our great scientist that made it possible, and we have had the opportunity to keep progressing through technology."

"Now I understand, Alckamus, how you managed to survive on our planet. I have another question I want to know based on your religion. Is there any being that was in paradise, the realm of which Helios comes from that one of his help betrayed him and has his own realm?"

"Well, that would be Hanos. He was an Anaku."

"Can you describe an

Anaku?"

"Yes, I can describe an Anaku for you, Alex. It is a being that is created for a purpose that only Helios knows. They're typically about five to seven feet tall with wings."

"Wait a minute. What you described to me

was an angel. So how does dream master come into play?"

"Well, he was an Anaku that sided with Hanos, and together, they were banned with other Anakus to their own realm."

"This is really weird, Alckamus. Our religions are the

exactly the same story, just different names and terms for things. Well, I see that we need to get back to our tasks. We already spent too much time on our religions, although I thought it was very interesting to compare the two. Now it's time to divert our attention to a bigger cause. That is to create an army resistance in our favor. I am putting my full trust in you, Alckamus. How are we going to convert General Dex?"

"Well, he does not respond well to threats. We will send him a message via hologram communication network, indicating that he needs to meet with me in a location he is familiar with. He will automatically assume I am Robert."

"How will he think that, Alckamus?"

"Well, you forget I can shape-shift? Changing into Robert won't be difficult at all."

Alckamus shape-shifted into Robert, indicating it was urgent to speak with him at New City Dr inside the park.

"Wow, that's cool. You look just like Robert. How long is the video?"

"It's about a thirty-second video. Are you

going to send it to Dex?"

"Right now."

Message had been sent, and he received a direct response from General Dex. "He had agreed to meet us in one hour."

"Alckamus, the hour has passed by. What is going on, guy?"

"I mean didn't he say he would be coming? I hear something. Do hide, Alex, now.

In a silent voice, Alex replied, "Okay, I'm hiding. Robert, Steven, and General Dex are here."

"It's Dex, Alckamus. Why are you here?"

"I sent for you. I have

a proposition for you. Join us in the resistance or join Robert and Steven? Think fast."

"But why should I help you?"

"Because if you don't, I will personally mail this conversation that is being recorded as we speak to your current bosses."

"Hold on! Give me a few minutes."

Meanwhile, back at the ship of Robert and Steven, "Hey, Steven, we need to send out a warrant for the arrest of Alckamus, and Steven, I hear they're getting pretty close now. I guess we can call them boyfriends. Ha-ha! General Dex, take care of them."

"Tell him to kill them on sight as soon as he is within range of them."

"I'm trying to notify him, but he is not responding. I wonder if something's going on with him. That's not like him to disappear."

Meanwhile, back at the location, where General Dex met with Alckamus, "What do you want, Alckamus?"

"Simple, General Dex. You be on my side and bring a lot of soldiers with you."

"Let's say I follow this.

How will it work? Make it fast because I just missed two messages from Robert and Steven, and they are going to drill me with questions."

"Okay, simple, Dex. Keep me posted at all times all the soldiers that are willing to join us. Hide their coming to this location when we let you know."

"Do you agree I know you're not happy with the current ruler?

"Let's take back control, and we will make this city prosperous once more like the glory days. Have you made your decision, General

Dex?"

"Yes, I did."

General Dex pulled out his version of a gun that melt anything with a single blast. Alckamus had his hands up in the air while letting Dex.

"Robert, Steven, pick up. I have Alckamus and Alex here, waiting for your order, sir."

"Kill them both right now."

While Dex looked away, Alckamus blasted him using the plasma beam coming from his eye.

"I told you, old friend, that this wasn't the way."

General Dex lay on the ground with half his body on the ground, bleeding green blood on the floor.

“I have one thing that I might say before I die, Alckamus. Get those bastards.”

“I will, old friend. You fought with honor, old friend. Rest in peace.”

Alex walked over to Alckamus and told him, "Come on, he is already dead."

"Now what is the next plan? I have no idea. What's next? Now that General Dex has been killed."

"Alex, fear not, my friend, for I have another solution to

this problem. Private Grunt. I knew him. General Dex would tell me how he would have posters of me and how much he idolized me."

"Well, can we keep him alive at least so that we could fully execute our goal, that is to eliminate Robert and Steven as rulers of new city. Please I want you back in power now to restore your city the way it was ran before."

"Correction, Alex, not the same way. I will make it better."

"This is what you have to let the public know."

“I will, but there are a lot of similarities between your planet and ours back home where I live. We have cops. Here they have cops as well, plus the military. So yeah, there are a lot of similarities between your planet and ours.”

“So, Alckamus, how

do the female give birth while all aliens on this planet are asexual creatures and how are big are the alien babies?"

"Well, Alex, I see you're quite bright. Well, our babies are born at least five feet tall."

"Holy shit, that's

amazing. I never would have imagined seeing a baby that size. That would really freak me out. A regular human is averaged to be about that height. Wow, amazing! I don't know what to think of what you just told me."

"Let me enlighten you,

Mr. Alex."

"Go ahead, Alckamus, enlighten me. I thought you already did by letting me know that your average baby's normal size happens to be five feet and zero inches. Amazing!"

"Well, there's more. Our plan was to

destroy your race and have our race, or as you would call us species live in your planet, but before that could happen, we needed to destroy your kind through diseases, plagues, wars, and drugs. All the above I have mentioned brings forth death for your people. Am I correct?"

"Yes, you're correct, Alckamus, but our species are very scared of your kind in the fact that you guys strongly believe in your god, and I've seen great things come out, believing. So in that case, we were more intimidated by your kind. Wow, again you never cease to amaze me, Alckamus. Here I am thinking how scared I was about aliens, but your kind are equally intimidated by us."

"That is why your government has been given instructions on implanting their whole population of citizens with a chip that they think is for their own good, but in reality, it's a way for us to have full control of the human race. Through technology and only technology, we will be able to keep the population at a certain number, and we can have full control. That's the way I was thinking it was a wrong way to think you may not trust me and you have no reason to, but I need to make things right. I never had someone guide me the right way. I thought I knew the right way. Luckily you came along and woke me up. I was a very selfish being with a black and gold heart."

"Well, Alckamus, you're wrong. We had our differences. We grew from them. There is no need to keep up with something that already happened. Please let's bury that hatchet and move on."

"Humans always move

on."

"Well most of us."

"You're right, Alex. I will from now try that in my own life. You're very wise, Alex. Who taught you how to be wise?"

"I don't know. I guess I have an old soul. Now, Alckamus, back to business. We have to work on getting Private Grunt in our pocket. How can we get this done without murdering him. Please we have to think logically."

"I agree, Alex. You're right."

"I know I am, Alckamus."

"Tell me the right

course of action to approach this matter, Alex."

"Well, how long it might take?"

"Be very patient. We have no other choice but to persuade Private Grunt."

"If you want I can shape-shift into an alien so that I can

persuade Private Grunt."

"Agreed, Alckamus."

"Agreed, Mr. Great."

"It's Alex to you, Alckamus. So get in contact with Private Grunt. I will make this happen for us."

Alckamus got Private Grunt to meet with

him at a secure location to discuss an urgent matter. Grunt arrived ten minutes later.

"I'm here to speak with Mr. Alex the Great. Are you Alex?"

"I'm here in need of your help Mr. Grunt. It seems to me that your alien can help me put a stop to Robert and Steven."

"Now how do I do that, Mr. Alex?"

"Well, simple by getting a resistance to stop them. Have our own army by recruiting your fellow soldiers. We need you, Grunt, and New City can't withstand any more negative ruling. Even Alckamus has vowed to make a change for the better. Come on, tell me you will help us."

"Okay, I will help both you guys. Robert and Steven are out of their fuckin' minds. They are so twisted and devious. I believe without a shadow of a doubt that they are going to try to kill our population."

"Wow. Mr. Grunt, that sounds so familiar. Our population is being killed off for years. Well, no worries, Mr. Grunt. You will have the backing of Alckamus and me at your disposal. So I advise you to start getting the resistance ready. Start now. Bring as many members as you can. The more, the merrier. Every help counts."

"Thanks, Alex. I will start immediately."

Private Grunt advised his team about the

crisis that prevailed in New City and their ruler.

"All I want to know is who will be fighting with me or against me. We need Alckamus back as our ruler. He himself has spoken regarding the changes that will become as a result of his new leadership. He will set forth new laws like a democracy. People will have to vote, and your ruler will be chosen by the civilians. Are you with me?"

There was a moment of silence, then . . .

"We will help you, Grunt."

"Thanks, guys."

Most of the soldiers joined Alex and Alckamus in the resistance movement.

“Okay, listen up. We will meet with Alckamus tonight and take back our city.”

“Yes, sir, Mr. Grunt.”

Grunt ran back to Alckamus and told him that most

everybody joined in.

“Okay, Mr. Grunt, thanks. Now gather your team. You’re their general now. So lead them with a perfect strategy.”

“I will, Alckamus.”

An hour had passed, and Private Grunt brought in his team.

"Okay, everybody in formation now."

As everyone lined up, Alckamus came to speak to them.

"Everybody, listen up. I am very pleased for everyone's bravery and courage. This won't be forgotten by me, and when I'm restored to power, each and every one of you will be heavily compensated as your rewards for helping me getting back to my throne where I truthfully belong. So that being said, I'm going to let General Grunt take control of the battlefield."

"Thanks, Alckamus.

I'm ready. I want five privates circling the ship. Since they think we are still okay, we will be catching them by surprise, so let's go and get back control of our city."

Alex, Alckamus, and Private Grunt, and all his team headed over to the ship that Robert and Steven were occupying.

“Everybody, get in position.”

“Yes, sir, General Grunt.”

“Grunt, get in touch with Robert and Steven. Tell them that they need to come up.”

“Okay.”

He called Robert and Steven through a screen cellphone device.

“Hello, Robert, I need to speak with you. Could I come up?”

“Sure, Grunt, come on up.”

As Grunt went up

with five men, he was stopped by guards.

"Men, it's okay. This is Private Grunt, our best soldier we have here. Mr. Grunt, how can I help you?"

"Could we speak in your office?"

"Sure, but they have to stay here."

“Could I bring at least someone with me to keep me company?”

“Oh okay.”

When they were in Robert and Steven’s office, the resistance team was getting in position to hear the command word.

“So Mr. Grunt, I noticed you have been

disappearing from time to time. Where do you go?"

"Well, I like to take time to relax. My mind needs to unwind right now."

The team came in, eliminated all the soldiers, and stormed in with Alckamus, but there was one problem. Grunt was being held in a position of death. If Robert or Steven pressed one button, the whole ship would blow up.

At this moment, Alckamus pointed his eye plasma beam laser toward Robert, who had his finger on the button.

“Don’t make me do it, Alckamus. I will kill all of us. All we were doing was what you have done for years, enslaving your race, the way our race was enslaved all these years.”

“Well, Robert, I got one word for ya. Bye.”

Alckamus blasted him with his plasma power, causing Robert to explode. Now Steven tried to press the button and ran toward the machine, and Alckamus blasted him with his plasma, exploding Steven as well.

“Thanks, Alex and

everyone here, for assisting me today in becoming your leader, but I just have one thing to take care of. Soldiers, throw Alex in a cell. Ha-ha! General Grunt, you are now my assistant. Make sure you bring the media and every outlet so I can announce I am New City's ruler again."

About the Author

Jose M. Vasquez lives in Brooklyn with his loving family. He is a true Brooklynite who has a creative vision of the arts. For more information, you may visit the website Jose M. Vasquez.com

www.ingramcontent.com/pod-product-compliance
Lightning Source LLC
Chambersburg PA
CBHW071351170526
45165CB00001B/3